Alexander G. Ramm

Iterative Methods for Calculating Static Fields and Wave Scattering by Small Bodies

Springer-Verlag
New York Heidelberg Berlin

Alexander G. Ramm
Department of Mathematics
Kansas State University
Manhattan, Kansas 66506
U.S.A.

59 3271

AMS Classifications: 35L99, 35P25, 27A40, 76-02

Library of Congress Cataloging in Publication Data

Ramm, A. G. (Alexander G.)
 Iterative methods for calculating static fields and
wave scattering by small bodies.

 Bibliography: p.
 Includes index.
 1. Waves—Mathematics. 2. Scattering (Physics)—
Mathematics. 3. Electrostatics—Mathematics.
4. Iterative methods (Mathematics) I. Title.
QC157.R35   530.1'24   82-3266
ISBN 0-387-90682-7 (U.S.)   AACR2

Printed in the United States of America

9 8 7 6 5 4 3 2 1

ISBN 0-387-90682-7 Springer-Verlag New York Heidelberg Berlin
ISBN 3-540-90682-7 Springer-Verlag Berlin Heidelberg New York

To my parents and July

ABSTRACT

Wave scattering by small bodies of arbitrary shape is of interest in geophysics, astrophysics, electrical engineering, colloidal chemistry, medicine, physics of the atmosphere and ocean, and other fields. The theory of wave scattering by small bodies is based on the theory of static boundary value problems. In this monograph iterative methods for solving exterior and interior static boundary value problems are given. Approximate analytical formulas for static fields and functionals of these fields, e.g., capacitances and polarizability tensors, are obtained. Two-sided estimates of these functionals are derived. Analytical approximate formulas for the scattering matrix in the problem of acoustic and electromagnetic wave scattering by a single homogeneous body of arbitrary shape, and also for a single flaky-homogeneous body and for a system of r bodies, $r \sim 10$ and $r \sim 10^{23}$, are given. The dependence of the scattering matrix on the boundary conditions (the Dirichlet, Neumann, and impedance boundary conditions are discussed) is studied. Numerical examples and some applications are given. The stability of the iterative processes is analyzed. The book is self-contained and most of the results cannot be found in other books.

In order to understand this book a very modest knowledge of mathematical analysis is required (for most of the book only calculus is needed). Some of the problems in the applied sciences which require the theory presented include light propagation in colloidal solutions, light scattering by cosmic dust, radio wave propagation in hail and rain, determination of the size distribution of particles from the observed scattering data (e.g., size distribution of particles in smoke, fog, muddy water, etc.), calculation of the initial field in an aperture of an antenna from the observed field scattered by a probe placed in the aperture, etc.

The book is addressed to mathematicians, engineers, and physicists interested in the computation of static fields, and the theory of wave scattering by small bodies of arbitrary shape and its applications.

ACKNOWLEDGMENTS

I am thankful to the Air Force Office of Scientific Research (AFOSR) for its financial support, to ONR for its support, to the staff of Springer-Verlag for their generous help; and to Janet Vaughn for her expert typing.

Table of Contents

Preface

Iterative methods for calculating static fields are presented in this book. Static field boundary value problems are reduced to the boundary integral equations and these equations are solved by means of iterative processes. This is done for interior and exterior problems and for various boundary conditions. Most problems treated are three-dimensional, because for two-dimensional problems the specific and often powerful tool of conformal mapping is available. The iterative methods have some advantages over grid methods and, to a certain extent, variational methods:

(1) they give analytic approximate formulas for the field and for some functionals of the field of practical importance (such as capacitance and polarizability tensor),

(2) the formulas for the functionals can be used in a computer program for calculating these functionals for bodies of arbitrary shape,

(3) iterative methods are convenient for computers.

From a practical point of view the above methods reduce to the calculation of multiple integrals. Of special interest is the case of integrands with weak singularities. Some of the central results of the book are some analytic approximate formulas for scattering matrices for small bodies of arbitrary shape. These formulas answer many practical questions such as how does the scattering depend on the shape of the body or on the boundary conditions, how does one calculate the effective field in a medium consisting of many small particles, and many other questions. In particular these formulas allow one to solve the inverse radiation problem, which can be formulated as follows: If (E,H) is the field scattered by a small probe placed at the point x in an electromagnetic field

(E_0, H_0), how does one calculate $(E_0(x), H_0(x))$ from a knowledge of the scattered field (E, H)?

We also present two-sided variational estimates of capacitances and polarizability tensors. This book is based mostly on the author's papers and results. But the subject is classical and there have been many papers and books written on this subject. Some of them are cited in the bibliography but the bibliography is, of course, incomplete. On the other hand, the author has tried to make this book self-contained. References to formulas in a different section have the number of the section in front of the number of the formula. References to formulas in different chapters also contain the chapter number. The sign □ denotes the end of a proof. Theorem 6.1.1 means Theorem 1 in Chapter 6, §1. The number of chapter is omitted in references to the formulas of this chapter. Most of the sections are divided into subsections. A subject index is not included because the table of contents is sufficiently detailed and can be used instead. The bibliography serves also as an author index.

Introduction

The aim of this introduction is to give a brief review of the methods of calculating electrostatic fields, to mention some problems solved in closed form, and to formulate the objectives of this book.

Only the three-dimensional problems are discussed, because the complex-variable methods for two-dimensional problems have been widely discussed in the literature. The problems solvable in closed form are collected in [32,17,18,2,6,10,22]. The method of separation of variables has been used to solve the static problems for an ellipsoid and its limiting forms (disks, needles), for a half-plane, wedge, plane with an elliptical aperture, hyperboloid of revolution, parabaloid of revolution, cone, thin sperical shell, spherical segment, two conducting spheres, and some other problems [32,17,18]. Electrostatic fields in a flaky medium (with parallel and sectorial boundaries) have been studied [6,18]. Some of the problems were solved in closed form using integral equations, e.g., the problems for a disk, spherical shell, plane with a circular hole, etc. Wiener-Hopf, dual, and singular integral equations were used [18,6,25,37]. Electrostatic problems for a finite circular hollow cylinder (tube) were studied in [35] by numerical methods. The capacitance per unit length of the tube and the polarizability of the tube were calculated. The authors reduced the integral equation for the surface charge to an infinite system of linear algebraic equations and solved the truncated system on a computer. Their method depends heavily on the particular geometry of the problem and does not allow one to handle any local perturbations of the shape of the tube. In [21] the variational methods of Ritz, Trefftz, the Galerkin method, and the grid method are discussed in connection with the static problems. However, no specific properties of these problems are used. These methods are presented in a more general setting in [20,16]. In practice these methods are time-consuming and variational methods in three-dimensional static problems probably have some advantages over the grid method. There exists a vast literature on calculation of the capacitances of perfect conductors [10,27]. In [10] there is a reference section which gives the capacitances of conductors of certain shapes. In [27,26] a systematic exposition of variational methods for estimation of the capacitances and other functionals of practical interest is given. In [33] there are some programs for calculating the two-dimensional static fields using integral equations method.

In [30] some geometrical properties of the lines of electrical field strength are used for approximate calculations of the field. This approach is empirical.

The objective of this book is to present systematically the use of integral equations for calculating the static fields and two functionals of these fields, the capacitance and polarizability tensor for bodies of arbitrary shape. The method gives approximate analytical formulas for calculations. These formulas can be used to construct a computer program for calculating the capacitance and polarizability tensor. The case of several bodies is also discussed as well as the case of a flaky-homogeneous body, e.g., a coated particle. Two-sided variational estimates of capacitances and polarizability tensors are given. The case of thin unclosed metallic screens is considered as well as the case of perfect magnetic films. This latter problem of calculating of the magnetic polarizability of an ideal magnetic film is of importance because such films are used as memory elements of computers. The above-mentioned formulas for capacitances and polarizability tensors allow one to give approximate analytical formulas for the scattering matrix in the problem of wave scattering by small bodies of arbitrary shape. This is done for scalar and electromagnetic waves. The dependence of the scattering matrix on the boundary conditions on the surface of the scatterer is investigated. Furthermore, the wave scattering in a medium consisting of many small particles is studied and equations for the effective field in the medium are derived. This makes it possible to discuss the inverse problem of determining the properties of such a medium from a knowledge of the waves scattered by this medium.

The theory of wave scattering by small bodies was originated by Rayleigh (1871), who studied various aspects of this theory until his death (1919). During the last century many papers were published in this field but for the first time analytical approximate formulas for the polarizability tensors and scattering matrix were found in [28a,b,j,p] and summarized in the monograph [28u].

Here these and other results are presented systematically. The author hopes that these results can be used by engineers, physicists, and persons interested in atmospheric and ocean sciences, radiophysics, and colloidal chemistry. Radio wave scattering by rain and hail; light scattering by cosmic dust, muddy water, and colloidal solutions; methods of nondestructive control; and radiomeasurement technique are just a few examples of possible applications of the theory of wave scattering by small bodies of arbitrary shape.

The structure of the book is clear from the contents. The author hopes that only a very modest knowledge of mathematics is required from the reader. Although there are some new mathematical results in the book, as a whole the book is addressed to an audience which applies mathematics. Therefore some of the results are not formulated as theorems. Only Chapter 6 is of purely mathematical nature.

Chapter 1. Basic Problems

§1. Statement of the Electrostatic Problems for Perfect Conductors

1. The basic equations of electrostatics are well known [17]:

$$\text{curl } E = 0, \quad \text{div } D = \rho, \quad D = \varepsilon E, \tag{1}$$

where E is the electric field, D is the induction, $\rho(x)$ is the charge distribution, and ε is the dielectric constant of the medium. If the medium is homogeneous and isotropic, then ε is constant; if it is isotropic but nonhomogeneous, then $\varepsilon = \varepsilon(x)$, $x = (x_1, x_2, x_3)$. In the general case $\varepsilon = \varepsilon_{ij}(x)$, $1 \leq i, j \leq 3$, is a tensor. The boundary condition on the surface Γ of a conductor is of the form

$$E_t\big|_\Gamma = N \times E\big|_\Gamma = 0, \tag{2}$$

where N is the unit outer normal to Γ. If σ is the surface charge distribution then

$$D_N = (D, N) = \sigma. \tag{3}$$

The vectors E and D are to be finite and can have discontinuities only on the surfaces of discontinuity of $\varepsilon(x)$, i.e., on the surfaces which are the boundaries of domains with different electrical properties (interface surfaces). The boundary conditions on such surfaces are

$$E_{1t} = E_{2t}, \quad D_{1N} = D_{2N}, \text{ respectively} \tag{4}$$

where 1 and 2 stand for the first and second medium, respectively. A perfect conductor in electrostatics is a body with $\varepsilon = +\infty$. Let us define an insulator in electrostatics as a body with $\varepsilon = 0$, i.e., on its surface

$$D_N\Big|_\Gamma = 0. \tag{5}$$

This definition is useful because a superconductor behaves in a magnetic field H like an insulator in the electric field $E = H$. Indeed, on the surface of the superconductor the boundary condition

$$B_N\Big|_\Gamma = 0 \tag{6}$$

holds, where B is the magnetic induction [17].

2. Many problems of practical interest in quasistatic electrodynamics can be reduced to static problems.

For example let a conductor Ω be placed in a harmonic electromagnetic field. Let the wave length λ of the field be much larger than the characteristic dimension a of Ω, $\lambda \gg a$. In practice $\lambda > 0.2a$ is often enough. If the depth δ of the skin layer is small, $\delta \ll a$, then the calculation of the field scattered by this body can be reduced to the static problem

$$\text{div } B = 0, \quad \text{curl } B = 0 \quad \text{in } \Omega_e, \tag{7}$$

$$B_N\Big|_\Gamma = -B_{0N}\Big|_\Gamma, \quad B(\infty) = 0. \tag{8}$$

Here Ω_e is the exterior of the domain Ω, B_0 is the magnetic induction at the location of Ω. One can assume that B_0 is constant since $a \ll \lambda$, i.e., the exterior field does not change significantly within the distance a. It is clear that the problem of (7)-(8) is equivalent (formally) to the problem of the insulator in the exterior electrostatic field $E_0 = B_0$.

It is worthwhile to mention that many problems of thermostatics, hydrodynamics, and elastostatics can be reduced to static problems similar to the above.

3. Let us formulate two basic problems of electrostatics.
Problem A. A conductor is placed in a given electrostatic field. Find the charge distribution σ induced on its surface.

Problem B. A conductor has total charge Q. Find the surface charge distribution σ.

Problem B'. A conductor is at potential V. Find σ.

In these problems the conductor may be a single body or a system of bodies.

4. In most books on electrostatics the third boundary condition is not discussed. Nevertheless some practical problems (such as the calculation of the resistance of linearly polarizable electrodes, and the calcu-

lation of the skin effect) can be reduced to the static boundary value
problem with the third boundary condition.

5. Let us formulate the basic problems of electrostatics as problems
of the potential theory. Let ε be a constant. Then from (1) it follows
that

$$E = -\nabla\phi, \qquad \Delta\phi = -\rho. \tag{9}$$

In the domain free from charge one has

$$\Delta\phi = 0. \tag{10}$$

If the given exterior field is $E_0 = -\nabla\phi_0$, then

$$\phi = \phi_0 + v \tag{11}$$

and v satisfies (10). The boundary condition (2) takes the form

$$\phi\big|_\Gamma = \text{const}, \tag{12}$$

while (3) takes the form

$$-\varepsilon \frac{\partial\phi}{\partial N}\bigg|_\Gamma = \sigma. \tag{13}$$

Problem A can be formulated as follows:

Find the solution ϕ of (10) of the form (11), subject to condition
(12), such that

$$v(\infty) = 0 \quad \text{and} \quad \int_\Gamma \left(\frac{\partial\phi}{\partial N}\right) ds = 0. \tag{14}$$

This condition (14) means that the total surface charge on the conductor
is zero (electroneutrality). Since $\int_\Gamma (\partial\phi_0/\partial N) ds = 0$, condition (14) can
be rewritten as

$$\int_\Gamma \left(\frac{\partial v}{\partial N}\right) ds = 0. \tag{15}$$

Problem B can be formulated as follows:

Find the solution ϕ of (10) subject to (12) and such that

$$-\varepsilon \int_\Gamma \left(\frac{\partial\phi}{\partial N}\right) ds = Q, \qquad \phi(\infty) = 0. \tag{16}$$

The constant in condition (12) should be found in the process of solving
problems A and B. It has physical meaning as the potential of the conductor. It is well known and very easy to prove that problems A and B
have unique solutions.

6. If the conductor is a thin unclosed metallic screen, then the edge condition must be satisfied. Let F denote the screen and L denote its edge. Then the edge condition can be written as

$$|\phi(x)| \sim \{g(x)\}^{\frac{1}{2}}, \qquad g(x) \equiv \min_{t \in L} |x-t|. \tag{17}$$

The function $g(x)$ is the distance from the point x to the edge. From (17) it follows that

$$|E| = |-\nabla\phi| \sim \{g(x)\}^{-\frac{1}{2}}, \qquad \sigma(s) \sim \{g(s)\}^{-\frac{1}{2}}, \tag{18}$$

where $s \in F$. Condition (17) is easy to understand if one notes that the potential near the edge of the wedge behaves like $r^\nu \sin(\nu\theta)$, where (r,θ) are the polar coordinates, $\nu = (2 - \theta_0 \pi^{-1})^{-1}$, and θ_0 is the angle of the wedge. If $\theta_0 = 0$ (this is the case of the screen) then $\nu = 0.5$ and one obtains (17).

§2. Statement of the Basic Problem for Dielectric Bodies

1. Let a dielectric body Ω with dielectric constant ε_i placed in a medium with dielectric constant ε_e. The basic electrostatic problem is to find the electric field which occurs if one places the body in the given electrostatic field $E_0 = -\nabla\phi_0$. This problem can be formulated as

$$\Delta\phi = 0 \quad \text{in} \quad \Omega \quad \text{and} \quad \Omega_e, \tag{1}$$

$$\varepsilon_i\left(\frac{\partial\phi}{\partial N}\right)_i = \varepsilon_e\left(\frac{\partial\phi}{\partial N}\right)_e \quad \text{on} \quad \Gamma, \tag{2}$$

$$\phi = \phi_0 + v, \qquad v(\infty) = 0. \tag{3}$$

Here and below $(\partial\phi/\partial N)_{i(e)}$ are the limit values of the normal derivatives from the interior (exterior) domains.

For v one has the problem

$$\Delta v = 0 \quad \text{in} \quad \Omega \quad \text{and} \quad \Omega_e, \tag{4}$$

$$\varepsilon_i\left(\frac{\partial v}{\partial N}\right)_i = \varepsilon_e\left(\frac{\partial v}{\partial N}\right)_e + (\varepsilon_e - \varepsilon_i)\left(\frac{\partial\phi_0}{\partial N}\right) \quad \text{on} \quad \Gamma, \qquad v(\infty) = 0. \tag{5}$$

If the body Ω is nonhomogeneous, then

$$\operatorname{div}(\varepsilon(x)\nabla\phi) = 0 \quad \text{in} \quad \Omega. \tag{6}$$

2. Let us give an example of a practical problem which leads to a boundary value problem with the third boundary condition

$$\left(\frac{\partial \phi}{\partial N} - h\phi\right)\Big|_\Gamma = f, \quad h = \text{const.} \tag{7}$$

Suppose that on the surface of a perfect conductor there is a thin film, e.g., an oxide film. Let ψ be the potential of the conductor and let ϕ be the potential of the exterior surface of the film. In electrochemistry it is assumed that $\phi - \psi$ is proportional to the current $j = -\gamma\nabla\phi$, where γ is the specific conductivity of the film. Therefore $\phi - \psi = b\gamma(\partial\phi/\partial N)$, where the constant b is the coefficient of proportionality. This condition is clearly of the form (7) with $h = (b\gamma)^{-1}$, $f = -h\psi$. The same condition will appear in the problem with an impedance surface or with a surface covered by a thin dielectric film.

In electrochemistry the surfaces of the metallic electrodes are not equipotential surfaces because of the electrochemical polarizations. The potential of the electrodes depends on the normal component of the electric current. If this dependence is linear one again reaches condition (7).

§3. Reduction of the Basic Problems to Fredholm's Integral Equations of the Second Kind

1. Let us state several formulas from potential theory which will be used below. Let

$$v(x) = \int_\Gamma \frac{\sigma(t)dt}{4\pi r_{xt}}, \quad w(x) = \int_\Gamma \frac{\partial}{\partial N_t} \frac{1}{4\pi r_{xt}} \mu(t)dt, \tag{1}$$

where $r_{xt} = |x-t|$ and N_t is the outer unit normal to Γ at the point t. Then

$$\left(\frac{\partial v}{\partial N}\right)_i^e = \frac{A\sigma \pm \sigma}{2}, \quad w_i^e = \frac{A^*\mu \mp \mu}{2}, \tag{2}$$

where

$$A\sigma = \int_\Gamma \sigma(t) \frac{\partial}{\partial N_s} \frac{1}{2\pi r_{st}} dt, \quad A^*\mu = \int_\Gamma \mu(t) \frac{\partial}{\partial N_t} \frac{1}{2\pi r_{st}} dt. \tag{3}$$

Note that

$$\Delta v = 0, \quad \Delta w = 0 \quad \text{in } \Omega \text{ and } \Omega_e \tag{4}$$

where Γ is the surface of Ω. Unless otherwise specified we assume that Γ is smooth. The formulas (1)-(4) are well known and can be found in every textbook on elliptic equations or equations of mathematical physics. For smooth surfaces the Liapunov-Tauber equality [7] holds:

$$\left(\frac{\partial w}{\partial N}\right)_i = \left(\frac{\partial w}{\partial N}\right)_e. \tag{5}$$

The above properties of the potential hold if the densities σ and μ are continuous. If the densities are Hölder continuous, the derivatives of the potentials have additional smoothness properties, which we do not state because they will not be used. A function f is called Hölder continuous if for some constants $c > 0$ and α, $0 < \alpha < 1$,

$$|f(t) - f(s)| \leq c|t - s|^\alpha.$$

2. In order to reduce Problem A from §1 to Fredholm's integral equation, let us look for a solution of this problem of the form

$$\phi = \phi_0 + \int_\Gamma \frac{\sigma(t)dt}{4\pi\varepsilon_e r_{xt}}. \tag{7}$$

The unknown function $\sigma(t)$ has a physical interpretation as the surface charge distribution. The function ϕ of (7) satisfies equation (1.10), condition (1.11), and the first condition in (1.14). Substitution of (7) into (1.13) with $\varepsilon = \varepsilon_e$, yields

$$\sigma = -A\sigma - 2\varepsilon_e \frac{\partial \phi_0}{\partial N} , \quad \int_\Gamma \sigma dt = 0. \tag{8}$$

The second equation is condition (1.15). If $\varepsilon \neq \varepsilon_e$ and the medium has dielectric constant ε, then

$$\sigma_\varepsilon = \frac{\varepsilon_e}{\varepsilon} \sigma \tag{9}$$

where σ_ε is the surface charge distribution in this new problem and σ is the solution of (8), i.e., the surface charge distribution in the original problem.

Exercise. Prove this statement.

It is well known [7] that every solution of the equation $\sigma = -A\sigma$ is of the form $\sigma = \text{const} \cdot \omega(t)$, where $\omega(t) \geq 0$, $\int_\Gamma \omega(t)dt > 0$. The function $\omega(t)$ describes the free charge distribution on the surface Γ of the conductor. Every solution of the adjoint equation $\mu = -A^*\mu$ is of the form $\mu = \text{const}$ [8]. From this and Fredholm's alternative it follows that problem (8) has a unique solution. Existence is guaranteed since $\int_\Gamma (\partial\phi_0/\partial N)ds = 0$, while uniqueness follows from the second condition in (8).

3. Let us look for a solution of Problem B of the form

$$\phi = \int_\Gamma \frac{\sigma(t)dt}{4\pi e_e r_{st}} \, , \tag{10}$$

where

$$\sigma = -A\sigma, \quad \int_\Gamma \sigma dt = Q. \tag{11}$$

Problem (11) has the unique solution

$$\sigma = Q\omega(s), \tag{12}$$

where $\omega(s)$ is the solution of (11) corresponding to $Q = 1$, i.e., an equilibrium charge distribution of total charge $Q = 1$ on the surface Γ of the conductor. It is easy to prove [8] that every solution of the first equation (11) is a constant multiple of $\omega(t)$.

4. Let us now consider the interior and exterior problems

$$\Delta\phi = 0 \quad \text{in} \quad \Omega, \quad \frac{\partial\phi}{\partial N} + h\phi\big|_\Gamma = f, \tag{13}$$

$$\Delta\phi = 0 \quad \text{in} \quad \Omega_e, \quad \frac{\partial\phi}{\partial N} - hu\big|_\Gamma = f, \tag{14}$$

where $h = h_1 + ih_2$, $h_1 \geq 0$, $h_2 \leq 0$, $|h| > 0$, $h = $ const. It is easy to prove that problems (13) and (14) have unique solutions. If one looks for a solution of the form $\phi = v$, where v is defined in (1), then the density σ of the potential v satisfies the equation

$$\sigma + T\sigma = -A\sigma + 2f \tag{15}$$

for problem (13), and

$$\sigma + T\sigma = A\sigma - 2f \tag{16}$$

for problem (14). Here A is defined in (3) and

$$T\sigma \equiv h \int_\Gamma \frac{\sigma dt}{4\pi r_{st}} \, . \tag{17}$$

For the Dirichlet problems

$$\Delta u = 0 \quad \text{in} \quad \Omega, \quad u\big|_\Gamma = f, \tag{18}$$

$$\Delta u = 0 \quad \text{in} \quad \Omega_e, \quad u\big|_\Gamma = f, \tag{19}$$

one looks for the solution of the form $u = w$, where w is defined in (1), and for μ obtains the equations

$$\mu = A^*\mu - 2f, \tag{20}$$

$$\mu = -A^*\mu + 2f, \tag{21}$$

respectively.

5. In order to reduce the basic problem of the electrostatics of dielectrics to integral equations let us look for a solution of the form (7). Using (2) and the boundary condition (2.2) one obtains the equation

$$\sigma = -\gamma A\sigma - 2\gamma\varepsilon_e \frac{\partial\phi_0}{\partial N}, \qquad \gamma = \frac{\varepsilon_i - \varepsilon_e}{\varepsilon_i + \varepsilon_e} \tag{22}$$

where ε_i is the dielectric constant of the body. If $\varepsilon_i = \infty$ then $\gamma = 1$. This is the case of a perfect conductor and in this case (22) is identical to (8). If $\varepsilon_i = 0$ then $\gamma = -1$. This is the case of an insulator and in this case (22) becomes

$$\sigma = A\sigma + 2\varepsilon_e \frac{\partial\phi_0}{\partial N} \tag{23}$$

6. If several conductors are placed in the exterior field $E_0 = -\nabla\phi$, then one looks for a potential of the form

$$\phi = \phi_0 + \sum_{j=1}^{p} \int_{\Gamma_j} \frac{\sigma_j(t)dt}{4\pi\varepsilon_e r_{xt}}. \tag{24}$$

From (24) and the boundary conditions

$$-\varepsilon_e \frac{\partial\phi}{\partial N}\bigg|_{\Gamma_m} = \sigma_m, \qquad 1 \leq m \leq p \tag{25}$$

one obtains the system of integral equations

$$\sigma_j(t_j) = - \sum_{m=1,m\neq j}^{p} T_{jm}\sigma_m - A_j\sigma_j - 2\varepsilon_e \frac{\partial\phi_0}{\partial N}, \qquad 1 \leq j \leq p, \tag{26}$$

where

$$T_{jm}\sigma_m = \int_{\Gamma_m} \frac{\partial}{\partial N_{t_j}} \frac{1}{2\pi r_{t_j t_m}} \sigma_m(t_m)dt_m, \tag{27}$$

$$A_j\sigma_j = \int_{\Gamma_j} \frac{\partial}{\partial N_{t_j}} \frac{1}{2\pi r_{t_j s_j}} \sigma_j(s_j)ds_j, \tag{28}$$

and the electroneutrality conditions should be satisfied

$$\int_{\Gamma_j} \sigma_j(t)dt = 0, \qquad 1 \leq j \leq p. \tag{29}$$

7. If several dielectric bodies are placed in the exterior field $E_0 = -\nabla\phi_0$, then the potential is of the form (24) and from the boundary conditions

$$\varepsilon_j\left(\frac{\partial\phi}{\partial N}\right)_i = \varepsilon_e\left(\frac{\partial\phi}{\partial N}\right)_e \quad \text{on} \quad \Gamma_j, \quad 1 \le j \le p, \tag{30}$$

one obtains the system of integral equations

$$\sigma_j(t_j) = -\kappa_j \sum_{m=1,m\neq j}^{p} T_{jm}\sigma_m - \kappa_j A_j \sigma_j - 2\kappa_j \varepsilon_e \frac{\partial\phi_0}{\partial N_{I_j}}, \tag{31}$$

where

$$\kappa_j = \frac{\varepsilon_j - \varepsilon_e}{\varepsilon_j + \varepsilon_e}, \quad 1 \le j \le p, \tag{32}$$

ε_j is the dielectric constant of the jth body, T_{jm} and A_j are defined in (27) and (28), and unless for some j_0 one has $\varepsilon_{j_0} = \infty$ there are no additional conditions on σ_j. Otherwise one should impose the electroneutrality condition (29) on σ_{j_0}.

8. Let us consider a flaky-homogeneous body placed in the exterior field $E_0 = -\nabla\phi_0$. Taking again the potential of the form (24) and using the boundary conditions

$$\varepsilon_j\left(\frac{\partial\phi}{\partial N}\right)_i = \varepsilon_{j-1}\left(\frac{\partial\phi}{\partial N}\right)_e \quad \text{on} \quad \Gamma_j \tag{33}$$

one finds the system of integral equations

$$\sigma_j(t_j) = -\gamma_j \sum_{m=1,m\neq j}^{p} T_{jm}\sigma_m - \gamma_j A_j \sigma_j - 2\gamma_j \varepsilon_e \frac{\partial\phi_0}{\partial N_{t_j}}, \tag{34}$$

where

$$\gamma_j = \frac{\varepsilon_j - \varepsilon_{j-1}}{\varepsilon_j + \varepsilon_{j-1}} \tag{35}$$

and T_{jm}, A_j are defined in (27), (28).

§4. Reduction of the Static Problems to Fredholm's Integral Equations of the First Kind

If the body is a metallic thin nonclosed screen it is not easy to reduce the static problems to a convenient Fredholm equation of the second kind. Some attempts to do this can be found in the literature [4]. Nevertheless it is easy to obtain Fredholm's integral equations of the first kind for the problem and to solve these equations by an iterative process.

Let us consider Problem A in §1. Looking for a potential of the form (3.7), using boundary condition (1.12) and denoting the constant potential on the surface of the conductor by V one gets

$$\int_\Gamma \frac{\sigma(t)\,dt}{4\pi\varepsilon_e r_{st}} = V - \phi_0.$$

(1)

The constant V is to be found from the condition

$$\int_\Gamma \sigma(t)\,dt = 0.$$

(2)

Problem B from §1 leads in a similar way to the equation

$$\int_\Gamma \frac{\sigma(t)\,dt}{4\pi\varepsilon_e r_{st}} = V,$$

(3)

which can be solved under the condition

$$\int_\Gamma \sigma\,dt = Q.$$

(4)

If $\eta(t)$ is the solution of (3) for $V = 1$, then the problem (3)-(4) has the solution

$$\sigma(t) = \frac{Q}{Q_1}\,\eta(t),$$

(5)

where

$$Q_1 = \int_\Gamma \eta(t)\,dt.$$

(6)

Problem B' in §1 is equivalent to equation (3) without additional conditions.

Chapter 2. Iterative Processes for Solving Fredholm's Integral Equations for the Static Problems

§1. An Iterative Process for Solving the Problem of Equilibrium Charge Distribution and Charge Distribution on a Conductor Placed in an Exterior Static Field

1. In §1.3, Problem A in §1.1 was reduced to the problem (1.3.8). It is well known [8] that the operator A in (1.3.8) is compact in $L^2(\Gamma)$ and in $C(\Gamma)$ provided that Γ is smooth (it is sufficient to assume that the equation of the surface in the local coordinates is $x_3 = f(x_1,x_2)$ and ∇f is Hölder continuous). It is also known [7] that $\lambda = -1$ is the smallest characteristic value of A which is simple. This means that $\lambda = -1$ is a simple pole of the resolvent $(A-\lambda I)^{-1}$ and the corresponding null space is one-dimensional, i.e., every solution of the equation $\sigma = -A\sigma$ is of the form $\sigma = \text{const} \cdot \omega(t)$, where $\omega(t)$ is the solution normalized by the condition $\int_\Gamma \omega dt = 1$. Let G_1 denote the null space of the operator $I + A^*$, where A^* is defined in (1.3.3). It is well known [7] and can be verified directly that $\mu = 1$ is a solution of the equation $\mu = -A^*\mu$. By the Fredholm alternative G_1 is one-dimensional. Let $G_1^\perp$ be the orthogonal complement to G_1 in $H = L^2(\Gamma)$. Then $G_1^\perp$ is the set of functions satisfying the condition $\int_\Gamma \sigma dt = 0$. If ϕ_0 is the electrostatic potential then $\int_\Gamma (\partial\phi_0/\partial N)dt = 0$. The theoretical basis for the iterative processes of this chapter is given in Chapter 6. In order to apply Theorem 1 from §6.1 it remains to check that equation $\sigma = -A\sigma$ has only the trivial solution in $G_1^\perp$. Every solution is of the form $\sigma = c\omega(t)$, $c = \text{const}$, $\int_\Gamma \omega(t)dt > 0$. Therefore $\int_\Gamma \sigma dt = 0$ implies that $c = 0$ and $\sigma = 0$. Theorem 6.1.1 and the above argument show that the following theorem holds.

Theorem 1. *Problem A in §1.1 has a unique solution* σ, *given by the iterative process*

$$\sigma_{n+1} = -A\sigma_n - 2\varepsilon_e \frac{\partial \phi_0}{\partial N}, \qquad \sigma_0 = -2\varepsilon_e \frac{\partial \phi_0}{\partial N}, \qquad \sigma = \lim_{n\to\infty} \sigma_n. \tag{1}$$

This process converges no more slowly than a geometric series with ratio q, $0 < q < 1$, *where* q *depends only on the shape of* Γ.

Remark 1. If Γ is a sphere then $q = 1/3$. The number $q = |\lambda_1 \lambda_2^{-1}|$, where $\lambda_1, \lambda_2, \lambda_3, \ldots$ are the characteristic numbers of A (i.e., $\phi_j = \lambda_j A \phi_j$ for some $\phi_j \neq 0$) numbered so that $|\lambda_1| \leq |\lambda_2| \leq |\lambda_3| \leq \ldots$, according to Theorem 6.1.1. One can calculate λ_1 and λ_2 numerically using methods given in [12], [16] and find q.

2. Let us solve Problem B by the iterative process given in Theorem 6.1.2. Problem B was reduced to problem (1.3.11). Its solution is of the form (1.3.12) and $\int_\Gamma \omega dt = 1$. Since $f(t) \equiv 1$ satisfies the condition $f_{G_1} \neq 0$ where f_{G_1} is the projection of f onto G_1 (note that G_1 is spanned by the function (1)) one can use Theorem 2 from §6.1. This yields

Theorem 2. *Problem B has a unique solution* σ, *given by the iterative process*

$$\sigma_{n+1} = -A\sigma_n, \qquad \sigma_0 = Q/S, \qquad \sigma = \lim_{n\to\infty} \sigma_n, \tag{2}$$

where $S = \text{meas } \Gamma$. *The process converges at the rate given in Theorem 1.*

Remark 2. It is easily seen that

$$\int_\Gamma \sigma_n dt = \int_\Gamma \sigma_{n-1} dt = \ldots = \int_\Gamma \sigma_0 dt = Q. \tag{3}$$

Indeed

$$-\int_\Gamma A\sigma dt = -\int_\Gamma \int_\Gamma \frac{\partial}{\partial N_t} \frac{1}{2\pi r_{st}} \sigma(s) ds dt = \int_\Gamma \sigma(s).$$

$$\left\{ \int_\Gamma - \frac{\partial}{\partial N_t} \frac{1}{2\pi r_{st}} dt \right\} ds = \int_\Gamma \sigma(s) ds$$

Here we used the known [8] formula

$$-\int_\Gamma \frac{\partial}{\partial N_t} \frac{1}{2\pi r_{st}} dt = 1.$$

This means that the iterative process (2) redistributes the fixed total charge on the surface approaching the equilibrium distribution.

3. Suppose Problem B is solved and $\omega(t)$ has been found. Then it is easy to solve Problem B'. Indeed, let V_0 be the potential of the conductor with $Q = 1$, i.e.,

$$\int_\Gamma \frac{\omega(t)}{4\pi\varepsilon_e r_{st}} = V_0.$$

Then the solution of Problem B' is

$$\sigma(t) = VV_0^{-1}\omega(t).$$

One can verify this directly.

Exercise. Do it.

§2. An Iterative Process for Solving the Problem of Dielectric Bodies in an Exterior Static Field

1. The problem is reduced in §1.3 to equation (1.3.22), where $-1 < \gamma < 1$, provided that $\varepsilon_i > 0$, $\varepsilon_e > 0$, $\varepsilon_i \neq 0$, and $\varepsilon_i \neq \infty$. It was already stated that all the characteristic values of A lie in the domain $|\lambda| \geq 1$. Therefore one can use Theorem 4 from §6.1. This implies the existence and uniqueness of solution of (1.3.22) and the convergence of the iterative process

$$\sigma_{n+1} = -\gamma A\sigma_n - 2\gamma\varepsilon_e \frac{\partial\phi_0}{\partial N}, \qquad \sigma_0 = \sigma_0; \qquad \sigma = \lim_{n\to\infty} \sigma_n \qquad (1)$$

where $\sigma_0 \in L^2(\Gamma)$ is arbitrary, to the solution of (1.3.22). The rate of convergence is that of the geometrical series with ratio q, $0 < q < |\gamma|^{-1}$. If $\sigma_0 = -2\gamma\varepsilon_e(\partial\phi_0/\partial N)$ then process (1) converges for $-1 \leq \gamma \leq 1$ and $q \leq |\lambda_2|^{-1}$, where λ_2 is the second characteristic value of A.

2. Suppose a flaky-homogeneous body described in §1.3 is placed in the exterior static field with the potential ϕ_0. The system of integral equations for this problem is (1.3.34).

Theorem 1. *The system (1.3.34) has a unique solution given by the iterative process*

$$\sigma_j = \lim_{n\to\infty} \sigma_j^{(n)},$$

$$\sigma_j^{(n+1)}(t_j) = -\gamma_j \sum_{m=1,m\neq j}^p T_{jm}\sigma_m^{(n)} - \gamma_j A_j\sigma_j^{(n)} - 2\gamma_j\varepsilon_e \frac{\partial\phi_0}{\partial N}\bigg|_{t_j}, \qquad (2)$$

$$\sigma_j^{(0)} = -2\gamma_j \varepsilon_e \frac{\partial \phi_0}{\partial N_{t_j}}, \qquad 1 \leq j \leq p, \tag{3}$$

which converges no more slowly than a geometric series with ratio q, $0 < q < 1$, where q depends only on the shapes of Γ_j.

Proof: Let us write (1.3.34) as

$$\sigma = -B\sigma + f, \tag{4}$$

where

$$\sigma = (\sigma_1, \ldots, \sigma_p), \quad f = \left(-2\varepsilon_e \gamma_1 \frac{\partial \phi_0}{\partial N_{t_1}}, \ldots, -2\varepsilon_e \gamma_p \frac{\partial \phi_0}{\partial N_{t_p}}\right),$$

B is the matrix operator of the form

$$B = \begin{pmatrix} \gamma_1 A_1 & \gamma_1 T_{12} & \cdots & \gamma_1 T_{1p} \\ \cdots\cdots\cdots\cdots\cdots\cdots\cdots \\ \gamma_p T_{p1} & \gamma_p T_{p2} & \cdots & \gamma_p A_p \end{pmatrix} \tag{5}$$

This operator acts in the space $H = L^2(\Gamma)$ of vector-valued functions with inner product

$$(\sigma, \omega) = \sum_{j=1}^{p} \int_{\Gamma_j} \sigma_j(t)\omega_j(t)dt. \tag{6}$$

In order to prove Theorem 1 it is sufficient to show that the equation

$$\sigma = -\lambda B\sigma \tag{7}$$

has only trivial solution for $|\lambda| \leq 1$ (see Theorem 4 from §6.1). Suppose $|\lambda| \leq 1$ and σ is a nontrivial solution of (7). Let us rewrite (7) as

$$\sigma_j = -\lambda\gamma_j\left(A_j\sigma_j + \sum_{m=1, m\neq j}^{p} T_{jm}\sigma_m\right). \tag{8}$$

If

$$v = \sum_{j=1}^{p} \int_{\Gamma_j} \frac{\sigma_j dt}{4\pi\varepsilon_e r_{xt}}$$

then

$$\left(\frac{\partial v}{\partial N_i} - \frac{\partial v}{\partial N_e}\right)\Bigg|_{\Gamma_j} = -\lambda\gamma_j \left(\frac{\partial v}{\partial N_i} + \frac{\partial v}{\partial N_e}\right)\Bigg|_{\Gamma_j}, \qquad 1 \leq j \leq p,$$

and

$$(1 + \lambda\gamma_j) \frac{\partial v}{\partial N_i} = (1 - \lambda\gamma_j) \frac{\partial v}{\partial N_e} \quad \text{on} \quad \Gamma_j, \quad 1 \leq j \leq p. \tag{9}$$

Let D_0 be the exterior domain with boundary Γ_1, let D_p be the interior domain with boundary Γ_p, and let D_j be the domain with boundary $\Gamma_j \cup \Gamma_{j+1}$. Let the a_j be arbitrary constants. Consider the identity

$$\sum_{j=0}^{p} a_j \int_{D_j} |\nabla v|^2 dx = \sum_{j=1}^{p} \int_{\Gamma_j} \bar{v}_j \left(a_j \frac{\partial v}{\partial N_i} - a_{j-1} \frac{\partial v}{\partial N_e} \right) ds. \tag{10}$$

From (9) and (10) it follows that

$$\sum_{j=0}^{p} a_j \int_{D_j} |\nabla v|^2 dx = \sum_{j=1}^{p} \int_{\Gamma_j} \bar{v} \left(a_j - a_{j-1} \frac{1+\lambda\gamma_j}{1-\lambda\gamma_j} \right) \frac{\partial v}{\partial N_i} ds. \tag{11}$$

If $|\gamma_j| < 1$ and $|\lambda| \leq 1$ then $|\lambda\gamma_j| < 1$. Let us set

$$a_0 = \varepsilon_e, \quad a_j = a_{j-1} \frac{1+\lambda\gamma_j}{1-\lambda\gamma_j}, \quad 1 \leq j \leq p.$$

Then (11) shows that $v \equiv 0$ and therefore $\sigma = 0$, i.e., $\sigma_j = 0$, $1 \leq j \leq p$. If $|\lambda| = 1$, $\lambda \neq 1$ then $\lambda\gamma_j \neq 1$ and the same argument shows that $\sigma = 0$. If $\lambda = 1$ and $\gamma_{j_0} = 1$ then $\varepsilon_{j_0} = \infty$ and $v|_{\Gamma_{j_0}} = \text{const}$. In this case one is interested in the potential in the domain exterior to Γ_{j_0} and has an identity similar to (11)

$$\sum_{j=0}^{j_0-1} a_j \int_{D_j} |\nabla v|^2 dx = \sum_{j=1}^{j_0-1} \int_{\Gamma_j} \bar{v} \left(a_j \frac{\partial v}{\partial N_i} - a_{j-1} \frac{\partial v}{\partial N_e} \right) ds + \int_{\Gamma_{j_0}} \bar{v} \, a_{j_0-1} \frac{\partial v}{\partial N_e} ds$$

$$= \sum_{j=1}^{j_0-1} \int_{\Gamma_j} \bar{v} \left(a_j - a_{j-1} \frac{1+\lambda\gamma_j}{1-\lambda\gamma_j} \right) \frac{\partial v}{\partial N_i} ds$$

$$- a_{j_0-1} \int_{\Gamma_{j_0}} \bar{v} \frac{\partial v}{\partial N_e} ds. \tag{12}$$

Because of the electroneutrality condition

$$\int_{\Gamma_{j_0}} \frac{\partial v}{\partial N_e} ds = 0, \tag{13}$$

and the boundary condition on the surface of the perfect conductor $v|_{\Gamma_{j_0}} = \text{const}$, the last integral in (12) vanishes. Therefore it follows from (12) that $\sigma \equiv 0$ provided that (13) holds. Note that (13) is equivalent to the equality

$$\int_{\Gamma_{j_0}} \sigma_{j_0} \, ds = 0. \tag{14}$$

Let us prove that $\lambda = -1$ is a semisimple characteristic value of the operator B. This will be important for construction of iterative methods of solution of equation (4). A characteristic number λ is called semi-simple if the equation $\sigma = \lambda B \sigma$ has nontrivial solutions and the equation $u = \lambda Bu + \sigma$ has no solution for any nonzero σ which is a solution of $\sigma = \lambda B\sigma$.

It can be proved that if B is compact then λ is semisimple if and only if it is a simple pole of the resolvent $(I - zB)^{-1}$ (see Chapter 6).

Suppose that

$$\sigma = -B\sigma, \quad \sigma \neq 0, \quad u = -Bu + \sigma. \tag{15}$$

Let $\int_{\Gamma_j} \sigma_j dt = Q_j$, $\int_{\Gamma_j} u_j dt = q_j$. Note that [8]

$$\int_{\Gamma} \frac{\partial}{\partial N_t} \frac{1}{2\pi r_{xt}} \, dt = \begin{cases} 0, & x \notin D, \\ -1, & x \in \Gamma, \\ -2, & x \in D, \end{cases} \tag{16}$$

where D is a bounded domain with a smooth boundary Γ. Integrating (15) over Γ yields

$$q_j = \gamma_j q_j + 2\gamma_j \sum_{m>j} q_m + Q_j, \quad j = 1, 2, \ldots, j_0, \tag{17}$$

because

$$\begin{aligned}
\int_{\Gamma_j} dt \, T_{jm}\sigma_m &= \int_{\Gamma_j} dt \int_{\Gamma_m} \frac{\partial}{\partial N_t} \frac{1}{2\pi r_{ts}} \sigma_m(s) ds \\
&= \int_{\Gamma_m} ds \, \sigma_m(s) \int_{\Gamma_j} dt \, \frac{\partial}{\partial N_t} \frac{1}{2\pi r_{ts}} = q_m \begin{cases} 0, & m < j \\ -2, & m > j \end{cases}.
\end{aligned} \tag{18}$$

Therefore (17) is a linear system with an upper triangular coefficient matrix. We have already showed that if $\int_\Gamma \sigma_{j_0} dt = Q_{j_0} = 0$ then $\sigma \equiv 0$. Since we assume that $\sigma \neq 0$ we have $Q_{j_0} \neq 0$. Since $\gamma_{j_0} = 1$ the last equation in (17) reads $q_{j_0} = q_{j_0} + Q_{j_0}$. Thus $Q_{j_0} = 0$ and $\sigma \equiv 0$. This contradicts the assumption that $\sigma \neq 0$. Therefore $\lambda = -1$ is a semisimple characteristic value of B.

The statement of Theorem 1 follows now from Theorem 6.1.1. Note that we need this theorem only in the case in which $\varepsilon_{j_0} = \infty$ because in this case -1 is the characteristic value of B. If each ε_j is finite then the operator B has no characteristic values in the unit disk $|\lambda| \leq 1$ and the iterative process (2) converges for any initial approximation, not necessarily satisfying the condition

$$\int_{\Gamma} f \, dt = 0. \tag{19}$$

This condition is satisfied by the initial approximation (3). □

3. Let us consider an iterative process for solving the problem of many bodies in the exterior static field.

In §1.3 this problem was reduced to system (1.3.31) in the case of dielectric bodies and to system (1.3.26) and conditions (1.3.29) for the case of perfect conductors. Since the case of perfect conductors can be treated as an instance of dielectric bodies with $\varepsilon_j = \infty$, let us consider system (1.3.31) and rewrite it as an operator equation

$$\sigma = -\tilde{B}\sigma + f, \tag{20}$$

where

$$\tilde{B}_{jm} = \kappa_j T_{jm}(1 - \delta_{jm}) + \kappa_j \delta_{jm}\Lambda_j, \qquad f_j = -2\kappa_j \varepsilon_e \frac{\partial \phi_0}{\partial N_{t_j}}, \tag{21}$$

and κ_j is defined in (1.3.32).

Theorem 2. *If* $|\kappa_j| < 1$, $1 \leq j \leq p$, *then equation* (20) *has a unique solution* σ *for any* $f \in H = L^2(\Gamma)$, *given by the iterative process*

$$\sigma_{n+1} = -\tilde{B}\sigma_n + f, \qquad \sigma = \lim_{n \to \infty} \sigma_n, \tag{22}$$

where $\sigma_0 \in H$ *is arbitrary. Process* (22) *converges no more slowly than a convergent geometric series. If* $\kappa_j = 1$ *for some* j *then equation* (20) *has a solution for any* $f \in H$ *such that*

$$\int_{\Gamma} f \, ds = 0. \tag{23}$$

This solution satisfies the condition

$$\int_{\Gamma} \sigma \, ds = 0. \tag{24}$$

There is only one solution of equation (20) *with* f *satisfying* (23) *in the class of functions* $\sigma \in H$ *satisfying* (24). *This solution can be found by the iterative process* (22) *where* σ_0 *satisfies condition* (24), *e.g.,* $\sigma_0 = f$. *The process converges at least as fast as a convergent geometric series.*

A proof of Theorem 2 is similar to the proof of Theorem 1 and can be left to the reader as an exercise. It can be found in [69].

§3. A Stable Iterative Process for Finding the Equilibrium Charge Distribution

The iterative process for solution of this problem is given in Theorem 2.1.2. However this process is unstable in the following sense. Consider the process with perturbations

$$\sigma_{n+1} = -A\sigma_n + \varepsilon_n, \quad ||\varepsilon_n|| \leq \varepsilon. \tag{1}$$

Since -1 is a characteristic value of A the operator $(I + A)^{-1}$ is not defined everywhere in H and the process (1) can diverge. For example, if $\varepsilon_n = f$, $||f|| < \varepsilon$, $\int_\Gamma f ds > 0$, and $\sigma_0 = f$, then process (1) diverges. Indeed, in this case $\sigma_n = \Sigma_{m=0}^n (-1)^m A^m f$. The Neumann series $\Sigma_{m=0}^\infty (-1)^m A^m f$ does not converge for elements $f \in N(I + A)$, where $N(I + A)$ is the null space of the operator $I + A$.

We have already seen that $\sigma \in N(I + A)$ has the property $\int_\Gamma \sigma dt \neq 0$. Therefore every f such that $\int_\Gamma f dt \neq 0$ can be represented as $f = c\sigma + f_1$, where $c = \text{const} \neq 0$ and $\int_\Gamma f_1 dt = 0$. Since -1 is a semisimple characteristic value the operator $(I + A)^{-1}$ is defined at f_1 and is not defined at σ. Hence $(I + A)^{-1}$ is not defined at f and σ_n does not converge as $n \to \infty$. One can verify this by a direct calculation using the identity

$$-\int_\Gamma A\sigma dt = \int_\Gamma \sigma dt \tag{2}$$

which is valid for any $\sigma \in H$ (see Remark 2.1.2). Integrating σ_n over Γ yields $\int_\Gamma \sigma_n dt = q(n+1)$, where $q = \int_\Gamma f dt \neq 0$. Therefore $\int_\Gamma \sigma_n dt \to \infty$ and σ_n does not converge in H. This simple argument gives the rate of divergence of the process (1).

This motivates the problem of constructing a stable iterative process for solving the problem (1.3.11). Let $Q = 1$ in (1.3.11), $S = \text{meas } \Gamma$, $\phi = S^{-1}$, $\omega = \phi + h$. Then from the equation $\omega = -A\omega$ it follows that

$$h = -Ah + F, \quad F = -\phi - A\phi, \quad \int_\Gamma \phi dt = 1. \tag{3}$$

Note that from (2) it follows that

$$\int_\Gamma F dt = 0. \tag{4}$$

The following theorem gives a stable iterative process for solution of (3).

This theorem is a particular case of the abstract Theorem 6.1.2.

Theorem 1. *The iterative process*

$$h_{n+1} = -Ah_n - \frac{1}{S} \int_{\Gamma} h_n dt + F, \qquad h_0 = F, \tag{5}$$

where F *is defined in* (3), *converges in* $H = L^2(\Gamma)$ *no more slowly than a convergent geometric series to an element* h, *and* $\omega = h + S^{-1}$ *is the unique solution of the problem* (1.3.11) *for* $Q = 1$. *Furthermore, the process* (5) *is stable: i.e., if*

$$g_{n+1} = -Ag_n - \frac{1}{S} \int_{\Gamma} g_n dt + F + \varepsilon_n, \qquad h_0 = F, \qquad |\varepsilon_n| \le \varepsilon, \tag{6}$$

then

$$\lim_{n \to \infty} \sup \|g_n - h\| = 0(\varepsilon). \tag{7}$$

Remark 1. Actually process (5) converges in $C(\Gamma)$ if Γ is smooth.

§4. An Iterative Process for Calculating the Equilibrium Charge Distribution on the Surface of a Screen

1. The basic equation (see §1.4) is

$$\int_{\Gamma} \frac{\eta(t)dt}{4\pi\varepsilon_e r_{st}} = 1. \tag{1}$$

Here Γ can be the surface of a metallic body or the surface of a metallic screen (an infinitely thin body). First consider the case of the solid body. Let

$$a(t) = \left\{ \int_{\Gamma} (4\pi\varepsilon_e r_{st})^{-1} dt \right\}^{-1}. \tag{2}$$

From the abstract results given in §6.4, the following theorem follows.

Theorem 1. *Let* $\eta_n = a(t)\psi_n$, *where* $a(t)$ *is defined in* (2),

$$\psi_{n+1} = (I - A_1)\psi_n + 1, \qquad \psi_0 = 1 \tag{3}$$

$$A_1\psi = \int_{\Gamma} (4\pi\varepsilon_e r_{st})^{-1} a(t)\psi(t)dt. \tag{4}$$

Then ψ_n *converges in* $H = L^2(\Gamma)$, *and* $\lim_{n \to \infty} \eta_n = \eta$ *is the solution of equation* (1).

Consider now the case in which Γ is the surface of a metallic screen. Let G be the edge of Γ,

$$h(t) = g^{-\frac{1}{2}}(t), \tag{5}$$

where $g(x)$ is defined in (1.1.17), and let

$$a_1(t) = h(t)\left\{\int_\Gamma \frac{h(s)ds}{4\pi\varepsilon_e r_{st}}\right\}^{-1} \tag{6}$$

Let $H_- = L^2(\Gamma; a_1^{-1}(t))$, where $L^2(\Gamma; p)$ is the L^2 space with the norm $||f||^2 = \int_\Gamma |f|^2 p dt$.

Theorem 2. *If $a(t)$ is replaced by $a_1(t)$ in Theorem 1, then the sequence η_n constructed in Theorem 1 converges in H_- to the solution of equation (1).*

2. Consider problem (1.4.1)-(1.4.2). If η is the solution of (1), then $V\eta$ is the solution of (1.4.1) with $\phi_0 = 0$. Let τ be the solution of the equation

$$\int_\Gamma \frac{\tau(s)ds}{4\pi\varepsilon_e r_{st}} = \phi_0. \tag{7}$$

This equation can be solved by the iterative processes given in Theorems 1 and 2. The constant V can be found from the condition (1.4.2),

$$V = \int_\Gamma \tau(t)dt\left(\int_\Gamma \eta(t)dt\right)^{-1}. \tag{8}$$

Let us summarize the above as a theorem.

Theorem 3. *The solution of problem (1.4.1)-(1.4.2) can be obtained by the formulas:*

$$\sigma = \lim_{n\to\infty} \sigma_n, \qquad \sigma_n = V_n\eta_n = \tau_n \tag{9}$$

where η_n is defined in Theorem 1 for the case of the volume conductor and in Theorem 2 for the case of the metallic screen, τ_n is defined by means of the iterative processes given in Theorems 1 and 2:

$$\tau_{n+1} = (I + A_1)\tau_n + \phi_0, \qquad \tau_0 = \phi_0, \tag{10}$$

$$V_n = \int_\Gamma \tau_n(t)dt\left(\int_\Gamma \eta_n dt\right)^{-1}. \tag{11}$$

Remark 1. It can be proved (see, e.g., [28u, Appendix 10]) that the operator $Tf = \int_\Gamma f(t)dt/(4\pi\varepsilon_e r_{st})$ maps $H_q(\Gamma)$ onto $H_{q+1}(\Gamma)$, where $H_q = W_2^q(\Gamma)$, $-\infty < q < \infty$, is the Hilbert scale of Sobolev spaces and $\Gamma \in C^\infty$ is a closed surface. In other words, T is a pseudodifferential elliptic operator of order -1.

Chapter 3. Calculating Electric Capacitance

§1. Capacitance of Solid Conductors and Screens

1. Suppose that the total charge of a conductor is Q and its potential is V. Then

$$Q = CV \tag{1}$$

and the coefficient C is called the capacitance of the conductor. If $\sigma(t)$ is the surface charge distribution, then

$$\int_\Gamma \frac{\sigma(t)dt}{4\pi\varepsilon_e r_{st}} = V, \quad s \in \Gamma, \tag{2}$$

and

$$\int_\Gamma \sigma dt = Q. \tag{3}$$

Thus

$$C = \int_\Gamma \sigma dt \left(\int_\Gamma \frac{\sigma dt}{4\pi\varepsilon_e r_{st}} \right)^{-1}. \tag{4}$$

The function $\sigma(t)$ can be calculated by the iterative processes given in §2.3 and §2.4. If σ_n is an approximation to σ then the potential

$$\int_\Gamma \frac{\sigma_n dt}{4\pi\varepsilon_e r_{st}} = V_n(s) \tag{5}$$

is not constant on Γ. In this case we introduce the averaged potential

$$\bar{V}_n = S^{-1} \int_\Gamma V_n(s)ds, \quad S = \text{meas } \Gamma. \tag{6}$$

If $\sigma_n \to \sigma$ in $H = L^2(\Gamma)$ then $\bar{V}_n \to V$ and

$$c^{(n)} = Q_n/V_n = \int_\Gamma \sigma_n dt \left(\frac{1}{S} \int_\Gamma ds \int_\Gamma \frac{\sigma_n(t)dt}{4\pi\varepsilon_e r_{st}}\right)^{-1}. \tag{7}$$

is an approximation to C. The iterative process (2.1.2) satisfies condition (2.1.3),

$$\int_\Gamma \sigma_n dt = Q, \quad n = 1,2,\dots . \tag{8}$$

In this case (7) can be written as

$$c^{(n)} = 4\pi\varepsilon_e S^2 \left(\int_\Gamma\int_\Gamma r_{st}^{-1}\delta_n(t)dtds\right)^{-1}, \tag{9}$$

where δ_n is the nth approximation to the solution of the problem

$$\delta = -A\delta, \quad \int_\Gamma \delta_n dt = S, \tag{10}$$

and A is defined as usual (see (1.3.3)). One can construct δ_n by means of the iterative process

$$\delta_{n+1} = -A\delta_n, \quad \delta_0 = 1. \tag{11}$$

Theorem 2.1.2 and (9) imply the following theorem.

Theorem 1. *Let*

$$c^{(n)} = 4\pi\varepsilon_e S^2 \left\{ \left(-\frac{1}{2\pi}\right)^n \int_\Gamma\int_\Gamma \frac{dtds}{r_{st}} \int_\Gamma \underbrace{\cdots}_{n} \int_\Gamma \psi(t,t_1)\cdots \right.$$
$$\left. \psi(t_{n-1},t_n)dt_1\cdots dt_n \right\}^{-1}, \tag{12}$$

where

$$\psi(t,s) = \frac{\partial}{\partial N_t} \frac{1}{r_{ts}}. \tag{13}$$

Then

$$|c - c^{(n)}| \le cq^n, \tag{14}$$

where $c > 0$ *and* $0 < q < 1$ *depend on the shape of the conductor and do not depend on* n. *The following inequality holds:*

$$4\pi\varepsilon_e S^2 J^{-1} = c^{(0)} \le C, \tag{15}$$

where

$$J = \int_\Gamma\int_\Gamma r_{st}^{-1} dsdt. \tag{16}$$

Proof: The first statement of Theorem 1 follows from Theorem 2.1.2, and the second statement will be proved in §3.3. □

Remark 1. Engineers used the following empirical method for calculation of capacitances for more than sixty years. They assumed that the
surface charge distribution of the total charge Q is constant, i.e.,
$\sigma = QS^{-1}$, calculated the averaged potential

$$V = S^{-1} \int_\Gamma ds \int_\Gamma \frac{QS^{-1}dt}{4\pi\epsilon_e r_{st}}$$

and found an approximation to C,

$$C \approx QV^{-1} = 4\pi\epsilon_e S^2 J^{-1}. \tag{17}$$

This is the zeroth approximation (12) about which Theorem 1 gives additional information: the inequality (15) and the way to improve the accuracy of the formula by passing to the nth approximation. Therefore
Theorem 1 gives a justification of the empirical rule described above.

Remark 2. One can use the iterative process given in §2.4 to calculate capacitances of conductors. Let η be the solution of equation (2)
with $V = 1$ and η_n be the approximation of the nth order to η. Then
$V_n \approx 1$ for large n and formula (7) takes the form

$$C_n \approx Q_n = \int_\Gamma \eta_n dt. \tag{18}$$

The subscript n in (18) indicates that C_n in (18) differs from $C^{(n)}$
in (12).

2. If the conductor is a thin metallic screen one can use formula
(18). The empirical method described in Remark 1, i.e., formula (17), is
not every accurate for screens. For example if the screen is a circular
disk the error in calculating the capacitance from formula (17) is 7.5%.

§2. Variational Principles and Two-Sided Estimates of Capacitance

1. Variational principles for capacitances have been widely discussed in the literature. The well-known book [27] should be mentioned
first. A reference book [10] on electrical capacitances is a collection
of numerical results and formulas for calculating of capacitance. The purpose of this chapter is to give some methods for deriving two-sided estimates for capacitance. Some of the results seem to be new (e.g., a necessary and sufficient condition for the Schwinger stationary principle to
be extremal and estimates of the capacitance of a conductor placed in a
nonhomogeneous dielectric medium).

2. We start with the following theorem.

Theorem 1. *Let* A *be a symmetric linear operator on a Hilbert space* H *with domain of definition* D(A). *The equality*

$$(Af,f) = \max_{\phi \in D(A)} \frac{|(Af,\phi)|^2}{(A\phi,\phi)} \tag{1}$$

holds if and only if A ≥ 0, *i.e.,* (Aφ,φ) ≥ 0 *for all* φ ∈ D(A). *By definition,* $|(Af,\phi)|^2/(A\phi,\phi) = 0$ *if* (Aφ,φ) = 0.

Remark 1. Let Af = g. In many physical problems (some examples will be given later) the quantity (f,g) has physical significance. J. Schwinger (see, e.g., [8]) used the stationary representation of this quantity

$$(f,g) = \operatorname{st}_{\phi \in D(A)} \frac{|(g,\phi)|^2}{(A\phi,\phi)}, \tag{2}$$

where st is the sign of the stationary value. In practice it is import-ant to know when this representation is extremal. Theorem 1 answers this question and provides a tool for deriving the lower bound for (Af,f).

Remark 2. For the equality

$$(Af,f) = \min_{\phi \in D(A)} \frac{|(Af,\phi)|^2}{(A\phi,\phi)} \tag{3}$$

to hold it is necessary and sufficient that A ≤ 0.

Proof of Theorem 1: If A ≥ 0 then $|(Af,\phi)|^2 \leq (Af,f)(A\phi,\phi)$ for all f, φ ∈ D(A). This is just the Cauchy inequality for the nonnegative bilinear form [f,φ] = (Af,φ). Hence $(Af,f) \geq |(Af,\phi)|^2/(A\phi,\phi)$ and equality holds for φ = λf, λ = const. If A ≤ 0 then -A ≥ 0 and

$$(-Af,f) = \max_{\phi \in D(A)} \frac{|(-Af,\phi)|^2}{(-A\phi,\phi)}. \tag{4}$$

Since max (-x) = -min x, where x is a real variable, one can see that (4) is equivalent to (3).

Let us prove the necessity of the condition A ≥ 0. Suppose that (Aψ,ψ) < 0 and (Aω,ω) > 0. Let ψ = ω + λψ, where λ is a real number, and (1) holds. Then

$$(Af,f) \geq \frac{|(Af,\omega)|^2 + 2\lambda \operatorname{Re}(Af,\omega)(Af,\psi) + \lambda^2|(Af,\psi)|^2}{(A\omega,\omega) + 2\lambda \operatorname{Re}(A\psi,\omega) + \lambda^2(A\psi,\psi)}. \tag{5}$$

Since $(A\omega, \omega)(A\psi, \psi) < 0$, the denominator of this fraction has two real zeros. Because the fraction is bounded from above the numerator has the same roots as the denominator. This implies that

$$\frac{|(Af, \omega)|^2}{|(Af, \psi)|^2} = \frac{(A\omega, \omega)}{(A\psi, \psi)} < 0, \tag{6}$$

which is a contradiction. Therefore $A \geq 0$ or $A \leq 0$. The case $A \leq 0$ is impossible. Indeed, in this case (1) implies that $(Af, f) \geq |(Af, \phi)|^2/(A\phi, \phi)$, i.e., $(Af, f)(A\phi, \phi) \leq |(Af, \phi)|^2$. Thus

$$(-Af, f)(-A\phi, \phi) \leq |(-Af, \phi)|^2, \tag{7}$$

which contradicts the Cauchy inequality for the nonnegative operator $-A$. Therefore $A \geq 0$. □

Remark 3. Let $A = A^*$. Then

$$(Af_i, f_j) = st \frac{(Af_i, \phi_j)(\phi_i, Af_j)}{(A\phi_i, \phi_j)}. \tag{8}$$

If $A \geq 0$ then for $i = j$ one can replace st by $\max$ in (8).

3. It is now easy to derive some lower bounds for capacitance. Let Γ be the surface of a perfect conductor which is charged up to the potential $V = 1$. If σ is the surface charge distribution then

$$A\sigma \equiv \int_\Gamma \frac{\sigma(t)dt}{4\pi\varepsilon_e r_{st}} = 1, \tag{9}$$

and

$$C = \int_\Gamma \sigma dt. \tag{10}$$

Since the integral operator A in (9) is selfadjoint and positive on $H = L^2(\Gamma)$, Theorem 1 says that

$$C = \max\left(\int_\Gamma \sigma(t)dt\right)^2 \left(\int_\Gamma \int_\Gamma \frac{\sigma(t)\sigma(s)dsdt}{4\pi\varepsilon_e r_{st}}\right)^{-1}, \tag{11}$$

where the maximum is taken over all $\sigma \in C(\Gamma)$ if Γ is a smooth closed surface. From (11) the well-known principle of Gauss [21] follows immediately:

$$C^{-1} = \min\left(Q^{-2} \int_\Gamma \sigma(t)u(t)dt\right). \tag{12}$$

This principle says that if the total charge Q is distributed on the surface Γ with the density $\sigma(t)$ and $u(t)$ is the potential of this charge distribution on Γ, then the minimal value of the right-hand side of (12) is C^{-1} and this minimal value is attained by the equilibrium charge distribution (i.e., by the solution of (9)).

From (11) it is easy to obtain some lower bounds for C. For example, if $\sigma = 1$ then (compare with (1.15))

$$C \geq C^{(0)} \equiv \frac{4\pi\varepsilon_e S^2}{J}, \qquad S = \text{meas } \Gamma, \qquad J = \int_\Gamma\!\int_\Gamma \frac{dsdt}{r_{st}} . \tag{13}$$

One can take

$$\sigma_m = \sum_{j=1}^m c_j \phi_j \tag{14}$$

where $\{\phi_j\}$ is a linearly independent system of functions in H and c_j are constants which are to be determined from the condition that the right-hand side of (11) is maximal. Then σ_m is an approximation to the equilibrium charge distribution and the value of the right-hand side of (11) is an approximation to C.

4. Let us formulate two classical variational principles for capacitance: the Dirichlet and Thomson principles [27]. The Dirichlet principle gives an upper bound for C. The Thomson principle is equivalent to the Gauss principle. Therefore combining the Dirichlet principle and (11) one can obtain two-sided estimates for C.

The *Thomson principle* is

$$C^{-1} = \min \int_{D_e} \varepsilon_e |E|^2 dx \tag{15}$$

where D_e is the exterior of the domain with boundary Γ, and the minimum is taken over the set of vector fields satisfying the conditions

$$\text{div } E = 0, \qquad \int_\Gamma (N, \varepsilon_e E) dt = 1, \tag{16}$$

where N is the outer unit normal to Γ at the point t. The minimum in (15) is attained at the vector $E = -\nabla u$, where

$$\Delta u = 0 \text{ in } D_e, \qquad u|_\Gamma = \text{const}, \qquad u(\infty) = 0, \qquad -\varepsilon_e \int_\Gamma \frac{\partial u}{\partial N} dt = 1. \tag{17}$$

The *Dirichlet principle* is

$$C = \min \int_{D_e} \varepsilon_e |\nabla u|^2 dx \tag{18}$$

where the minimum is taken over the set of functions $u \in C^1(D_e)$ such that

$$u\big|_{\Gamma} = 1, \quad u(\infty) = 0. \tag{19}$$

This minimum is attained at the function u which is the solution to the problem

$$\Delta u = 0 \quad \text{in} \quad D_e, \quad u\big|_{\Gamma} = 1, \quad u(\infty) = 0. \tag{20}$$

Both principles are particular cases of the principles formulated and proved in the next section.

5. If Γ is the surface of a screen the admissible functions in the variational principles should satisfy the edge condition: if L is the edge of Γ then

$$u \sim \{g(x)\}^{\frac{1}{2}}, \quad \sigma \sim \{g^{-\frac{1}{2}}(x)\}, \quad g(x) \equiv \min_{t \in L} |x-t|. \tag{21}$$

§3. Capacitance of Conductors in an Anisotropic and Nonhomogeneous Medium

Let $\varepsilon = \varepsilon_{ij}(x)$ be the tensor (a positive definite matrix) of dielectric permittivity of the medium and let D be a perfect conductor with a smooth boundary Γ. The problem of finding the capacitance of this conductor placed in the nonhomogeneous anistropic medium is of interest in many practical cases. For example, suppose a metallic body is placed partially in water. If the characteristic dimension of the conductor is small in comparison with the wavelength in the medium with large dielectric constant the capacitance determines the scattering amplitude. We assume for simplicity that $\varepsilon_{ij}(x) \in C^1(D_e)$ and $\varepsilon_{ij}(x) = \varepsilon_{ij}$ for sufficiently large x. This assumption guarantees that the basic results about existence of solutions to static problems are the same as for the Laplace operator (homogeneous medium). The principles A and B analogous to the Dirichlet and Thomson principles can be formulated as follows:

Principle A:

$$C = \min \int_{D_e} (\varepsilon \nabla u, \nabla u) dx \tag{1}$$

where the minimum is taken over the C^1 functions u(x) such that

$$u\big|_{\Gamma} = 1, \quad u(\infty) = 0. \tag{2}$$

In the statement of these principles the usual notations

$$(a,b) = \sum_{j=1}^{3} a_j b_j, \quad (\varepsilon a)_i = \sum_{j=1}^{3} \varepsilon_{ij}(x) a_j \tag{3}$$

are used.

Principle B:

$$c^{-1} = \min \int_{D_e} (\varepsilon E, E) dx, \qquad (4)$$

where the minimum is taken over the set of vector fields satisfying the conditions

$$\text{div } \varepsilon E = 0 \text{ in } D_e, \quad \int_\Gamma (N, \varepsilon E) dt = 1. \qquad (5)$$

Proof of Principle A: Assume that

$$\text{div}(\varepsilon \nabla u) = 0 \quad \text{in} \quad D_e \qquad (6)$$

and that (2) is valid. The Euler equation for the functional in (1) is (6). Therefore (6) and (2) are necessary conditions for the function which solves (1), (2). The solution of (6) and (2) exists and is unique. Let us show that the functional in (1) attains its minimum at this solution and this minimum is C. Let $\eta \in C^1(D_e)$ satisfies the conditions

$$\eta|_\Gamma = 0, \quad \eta(\infty) = 0. \qquad (7)$$

Then

$$\int_{D_e} (\varepsilon \nabla u + \varepsilon \nabla \eta, \nabla u + \nabla \eta) dx = \int_{D_e} (\varepsilon \nabla u, \nabla u) dx + \int_{D_e} (\varepsilon \nabla \eta, \nabla \eta) dx$$

$$+ 2\text{Re} \int_{D_e} (\varepsilon \nabla u, \nabla \eta) dx$$

$$\geq \int_{D_e} (\varepsilon \nabla u, \nabla u) dx. \qquad (8)$$

Here we took into consideration that ε is positive definite and

$$\int_{D_e} (\varepsilon \nabla u, \nabla \eta) dx = -\int_\Gamma (N, \eta \varepsilon \nabla u) dx - \int_{D_e} \eta \text{ div}(\varepsilon \nabla u) dx = 0. \qquad (9)$$

Furthermore,

$$\int_{D_e} (\varepsilon \nabla u, \nabla u) dx = -\int_\Gamma (N, \varepsilon \nabla u) u dt = \int_\Gamma (D, N) dt = Q, \qquad (10)$$

where D is the electrical induction. Therefore the minimum in (1) is equal to the capacitance C if u is the solution to problem (2), (6). □

Proof of Principle B: From (10) it follows that the right-hand side of (4) is equal to c^{-1} is $E = -A\nabla u$, where u is the solution to (2), (6) and the constant A is defined as

$$A = \left\{ -\int_\Gamma (N, \varepsilon \nabla u) dt \right\}^{-1} = Q^{-1}. \tag{11}$$

Let us show that any other E satisfying (5) gives a larger value to functional (4). Indeed,

$$\int_{D_e} (\varepsilon E + \varepsilon h,\ E+h) dx = \int_{D_e} (\varepsilon E, E) dx + \int_{D_e} (\varepsilon h, h) dx + 2 \operatorname{Re} \int_{D_e} (\varepsilon E, h) dx$$

$$\geq \int_{E_e} (\varepsilon E, E) dx. \tag{12}$$

Here the following identity was used

$$\int_{D_e} (\varepsilon E, h) dx = -A \int_{D_e} (\nabla u, \varepsilon h) dx$$

$$= A \int_{D_e} u\ \operatorname{div}(\varepsilon h) dx + A \int_\Gamma u(N, \varepsilon h) dt = 0. \quad \square \tag{13}$$

Remark 1. If

$$\varepsilon_{ij}(x) = \delta_{ij} = \begin{cases} 1, & i = j \\ 0, & i \neq j \end{cases},$$

then principles A and B are the Dirichlet and Thomson principles.

Remark 2. Principles A and B give estimates of the capacitance from above and from below.

Example 1. Let us take

$$E = -A\varepsilon^{-1} \nabla u \tag{14}$$

where ε^{-1} is the inverse matrix of ε, u is an arbitrary harmonic function in D_e (i.e., $\Delta u = 0$ in D_e) and

$$A^{-1} = -\int_\Gamma \frac{\partial u}{\partial N}\ dt. \tag{15}$$

Then condition (5) is satisfied. Let

$$u(x) = \frac{1}{S} \int_\Gamma \frac{dt}{4\pi r_{xt}}, \quad S = \operatorname{meas} \Gamma. \tag{16}$$

Then it is easy to see that the constant A defined in (15) is equal to 1. Therefore from (4) it follows that

$$C \geq 16\pi^2 S^2 \left\{ \int_{D_e} (\varepsilon^{-1} \nabla v, \nabla v) dx \right\}^{-1}, \tag{17}$$

where

$$v(x) \equiv \int_\Gamma r_{xt}^{-1} dt. \tag{18}$$

If $\varepsilon_{ij}(x) = \varepsilon_e \delta_{ij}$, i.e., the medium is isotropic and homogeneous, then (17) and Green's formula imply that

$$C \geq 16\pi^2 s^2 \varepsilon_e \left\{ -\int_\Gamma ds \left(\int_\Gamma \frac{dt}{r_{st}} \frac{\partial}{\partial N_s} \int_\Gamma \frac{dt}{r_{st}} \right) \right\}^{-1}. \tag{19}$$

Example 2. Let $\varepsilon_{ij}(x) = \varepsilon(x)\delta_{ij}$,

$$u(x) = |x|^{-1}, \qquad E = \frac{Ax}{|x|^3 \varepsilon(x)} \tag{20}$$

$$A = \left\{ \int_\Gamma \frac{(t,N)}{|t|^3} dt \right\}^{-1} = \frac{1}{4\pi}. \tag{21}$$

From (4) it follows that

$$C \geq 16\pi^2 \left\{ \int_{D_e} \frac{dx}{|x|^4 \varepsilon(x)} \right\}^{-1}. \tag{22}$$

In particular if $D_e = \{x: |x| \geq a\}$ and $\varepsilon(x) = \varepsilon(|x|) = \varepsilon(r)$, then

$$C \geq 16\pi^2 \left\{ 4\pi \int_a^\infty dr\, r^{-2} \varepsilon(r) \right\}^{-1} = 4\pi \left\{ \int_a^\infty dr\, r^{-2} \varepsilon(r) \right\}. \tag{23}$$

Actually, in this case C is equal to the right-hand side of (23) because (20) is the real electrostatic field corresponding to the equilibrium charge distribution on the sphere $r = a$ if $\varepsilon(x) = \varepsilon(r)$.

Example 3. Let all of the space be divided into n parts bounded by conical surfaces. Suppose that the jth cone cuts the solid angle ω_j on the unit sphere and the vertices of the cones are in the center of a metallic ball with radius a. Let the dielectric constant of the jth cone be $\varepsilon_0 \varepsilon_j(r)$. Then (22) says that

$$C \geq 16\pi^2 \varepsilon_0 \left\{ \sum_{j=1}^n \omega_j \int_a^\infty r^{-2} \varepsilon_j(r) dr \right\}^{-1}. \tag{24}$$

In particular, if $\omega_1 = \omega_2 = 2\pi$ then

$$C \geq 8\pi \varepsilon_0 \left\{ \int_a^\infty r^{-2} \varepsilon_1(r) dr + \int_a^\infty r^{-2} \varepsilon_2(r) dr \right\}^{-1}. \tag{25}$$

This example covers the case in which the ball is halfway immersed in the water.

It is clear from the above examples that principle B is easy to use in practice, with only the difficulty in the calculations. In application of principle A there is the additional difficulty of finding a set of functions which satisfy condition (2). If the surface Γ is a coordinate surface in some known coordinate system it is easy to find such functions and Principle A gives upper bounds on C. A more general situation is discussed in Example 5 below.

Example 4. Let us take Example 3 and substitute $u = a/r$ in (1). This yields

$$C \leq \epsilon_0 \sum_{j=1}^{n} \omega_j a^2 \int_a^\infty r^{-2} \epsilon_j(r) dr \qquad (26)$$

In particular, if $\omega_1 = \omega_2 = 2\pi$ one obtains

$$8\pi\epsilon_0 \left\{ \int_a^\infty r^{-2} [\epsilon_1(r) + \epsilon_2(r)] dr \right\}^{-1} \leq C \leq 2\pi\epsilon_0 a^2 \int_a^\infty r^{-2} \cdot$$
$$\cdot [\epsilon_1(r) + \epsilon_2(r)] dr, \qquad (27)$$

from (25) and (26). For $\epsilon_1(r) = \epsilon_2(r) = 1$, estimate (27) gives the exact value of C. One can improve the estimates taking more complicated admissible functions.

Example 5. Suppose that $r = F(\theta, \phi)$ is the equation of the surface of the conductor. Set $u = F(\theta, \phi)/|x|$ in (1). Then condition (2) holds and (1) yields the following upper bound on C:

$$C \leq \epsilon_e \int_0^\pi \int_0^{2\pi} \frac{d\theta d\phi \sin\theta}{F(\theta,\phi)} \{ F^2(\theta,\phi) + |F_\theta'(\theta,\phi)|^2 + \sin^{-2}\theta |F_\phi'(\theta,\phi)|^2 \}. \qquad (28)$$

§4. Physical Analogues of Capacitance

In heat transfer, electrodynamics of direct current, and other fields the mathematical formulation of the problems can be reduced to the solution of the Laplace equation. Therefore in these subjects there exist some quantities analogous to the capacitance.

For example heat conductance in a homogeneous medium can be defined as

$$G_T = \frac{k}{\epsilon} C, \qquad (1)$$

where k is the coefficient of thermal conductivity, ϵ is the dielectric constant, C is the electrical capacitance of the conductor, and G_T is the heat conductance of the body with the same shape.

If G_M is the magnetic conductance and μ is the magnetic permittivity then

$$G_M = \frac{\mu}{\varepsilon} C. \tag{2}$$

If G is the electric conductance and γ is the coefficient of electrical conductivity then

$$G = \frac{\gamma C}{\varepsilon}. \tag{3}$$

§5. Calculating the Potential Coefficients

1. Let n conductors be placed in a homogeneous medium with the dielectric permittivity $\varepsilon = 1$. Let Γ_j be the surface of the jth conductor. Because the equations of electrostatics are linear there is a linear dependence between the potentials V_j of the conductors and their total charges Q_j,

$$Q_j = \sum_{i=1}^{n} C_{ij} V_j, \quad 1 \le i \le n. \tag{1}$$

The coefficients C_{ij}, $i \ne j$ are called the electrical inductance coefficients and the coefficients C_{jj} are called the capacitance coefficients.

The quadratic form

$$U = \frac{1}{2} \sum_{i,j=1}^{n} C_{ij} V_j V_i \tag{2}$$

is the energy of the electrostatic field. Therefore this form is positive definite. It is well known that this is the case if and only if all the principal minors of the matrix C_{ij} are positive (Sylvester's criterion). In particular

$$C_{jj} > 0, \quad C_{jj} C_{ii} > C_{ij}^2, \quad \det(C_{ij}) > 0, \tag{3}$$

and

$$C_{ij} = C_{ji}, \quad 1 \le i,j \le n \tag{4}$$

since the matrix C_{ij} is real valued. We can rewrite (1) as

$$V_i = \sum_{j=1}^{n} C_{ij}^{(-1)} Q_j, \quad 1 \le i \le n. \tag{5}$$

The coefficients $c_{ij}^{(-1)}$ are called the potential coefficients. The following inequalities hold

$$c_{jj}^{(-1)} > 0, \quad c_{ij}^{(-1)} > 0; \quad c_{ij} < 0. \tag{6}$$

The first inequality in (6) holds because $c_{ij}^{(-1)}$ is a positive definite matrix if C_{ij} is. In order to prove the last inequality in (6) let us take $V_m = 0$ if $m \neq j$ and $V_j = 1$, then formula (1) shows that $Q_i = C_{ij}$. Therefore we must show that $Q_i < 0$. But $Q_i = -\varepsilon_e \int_{\Gamma_i} (\partial u/\partial N) ds$. Thus it is sufficient to prove that $(\partial u/\partial N)|_{\Gamma_i} \geq 0$. Here u is the electrostatic potential generated by the jth conductor, provided that the other conductors have zero potentials. The function u is a harmonic function (i.e., $\Delta u = 0$) and $u(\infty) = 0$, $u|_{\Gamma_j} = 1$. Since u is harmonic it cannot have extremal points inside the domain of definition. Therefore $0 < u < 1$ between the conductors.

Since $u|_{\Gamma_i} = 0$ according to our assumption, it is clear that $(\partial u/\partial N)|_{\Gamma_i} \geq 0$ and the last inequality in (6) is proved. The second inequality in (6) can be proved similarly.

2. The problem of determining the equilibrium charge distribution on the surfaces of a system of conductors can be reduced to the following system of integral equations (see (2.2.20) where κ_j should be replaced by 1 and $f = 0$):

$$\sigma = -\tilde{B}\sigma, \quad (\tilde{B}\sigma)_j = \sum_{\substack{m \neq j, m=1}}^{n} T_{jm}\sigma_m + A_j\sigma_j, \quad 1 \leq j \leq n,$$

$$\sigma = (\sigma_1, \ldots, \sigma_n), \tag{7}$$

$$\int_{\Gamma_j} \sigma_j dt = Q_j, \quad 1 \leq j \leq n. \tag{8}$$

Here Q_j is the total charge of the jth conductor. (See §2.2 and §2.3.)

Theorem 1. *The solution to problem (7)-(8) exists, is unique and can be found by the iterative process*

$$\sigma^{(k+1)} = -\tilde{B}\sigma^{(k)}, \quad \sigma_j^{(0)} = Q_j S_j^{-1}, \quad 1 \leq j \leq n, \quad S_j = \text{meas } \Gamma_j. \tag{9}$$

This theorem follows from Theorem 7.1.2.

Let us derive some approximate formulas for the potential coefficients. Taking $Q_j = \delta_{jm}$ in (5) yields

$$C_{im}^{(-1)} = V_i. \tag{10}$$

Let us substitute in the system of integral equations

$$\sum_{j=1}^{n} \int_{\Gamma_j} \frac{\sigma_j(t)dt}{4\pi\epsilon_r r_{tt_i}} = V_i, \quad 1 \le i \le n, \tag{11}$$

$\sigma_j^{(0)} = Q_j S_j^{-1}\delta_{jm}$ instead of $\sigma_j(t)$. Taking into account (1) one obtains

$$C_{im}^{(-1)} \approx \frac{1}{4\pi\epsilon_e S_m} \int_{\Gamma_m} \frac{dt}{r_{tt_i}}, \quad 1 \le i \le n. \tag{12}$$

The right-hand side of this formula is not constant on Γ_i because $\sigma_j^{(0)}$ is not the exact solution to (11). Therefore we take as an approximation to $C_{im}^{(-1)}$ the average of the right-hand side of (12). This yields

$$C_{im}^{(-1)} \approx c_{im}^{(-1)} \equiv \frac{1}{4\pi\epsilon_e S_m S_i} \int_{\Gamma_i} \int_{\Gamma_m} \frac{dsdt}{r_{st}}, \quad 1 \le i,m \le n. \tag{13}$$

One can improve formula (13) by using the higher order approximations to σ, say $\sigma^{(k)}$ defined in (9). In order to find some approximation to C_{ij} one can invert the matrix $c_{ij}^{(-1)}$, using the approximate values of $c_{ij}^{(-1)}$ given above.

3. Let us derive variational principles for the potential coefficients. To do so we take the potential energy of the electrostatic field

$$U = \frac{1}{2} \sum_{i,j=1}^{n} C_{ij}^{(-1)} Q_i Q_j \tag{14}$$

and set $Q_i = \delta_{im}$. This yields

$$2U = C_{mm}^{(-1)}. \tag{15}$$

Among various surface charge distributions such that

$$\int_{\Gamma_i} \sigma_i(t)dt = \delta_{im}, \quad 1 \le i \le n, \tag{16}$$

the distribution corresponding to the real electrostatic field minimizes U. Thus

$$C_{mm}^{(-1)} = \min \sum_{i,j=1}^{n} \int_{\Gamma_i} \int_{\Gamma_j} \frac{\sigma_i(t)\sigma_j(s)dsdt}{4\pi\epsilon_e r_{st}} \tag{17}$$

where the minimum is taken over the set of σ_j satisfying condition (16).

In order to derive a variational principle for C_{mm} we take $V_i = \delta_{im}$ in (2). This yields

$$2U = C_{mm}. \tag{18}$$

The energy of the electrostatic field with the potential $u(x)$ can be written as

$$U = \frac{1}{2} \int_{D_e} \varepsilon_e |\nabla u|^2 dx, \tag{19}$$

where D_e is the domain outside of the conductors. Let u satisfy the conditions

$$u|_{\Gamma_m} = 1, \quad u|_{\Gamma_i} = 0, \quad i \neq m, \quad u(\infty) = 0, \quad u \in C^1(D_e). \tag{20}$$

Then

$$C_{mm} = \min \int_{D_e} \varepsilon_e |\nabla u|^2 dx, \tag{21}$$

where the minimum is taken over the set of functions u satisfying condition (20).

Let $m \neq j$ and assume

$$\int_{\Gamma_i} \sigma_i dt = \delta_{ij} + \delta_{im}, \quad 1 \leq i \leq n. \tag{22}$$

From (14) and (22) it follows that

$$2U = C_{mm}^{(-1)} + 2C_{mj}^{(-1)} + C_{jj}^{(-1)}. \tag{23}$$

Therefore

$$C_{mm}^{(-1)} + 2C_{mj}^{(-1)} + C_{jj}^{(-1)} = \min \sum_{i,k=1}^{n} \int_{\Gamma_i} \int_{\Gamma_i} \frac{\sigma_i(t)\sigma_k(s)dsdt}{4\pi\varepsilon_e r_{st}} \tag{24}$$

where the minimum is taken over the set of functions σ_i satisfying condition (22).

If $C_{jj}^{(-1)}$, $1 \leq j \leq n$ are already calculated, then one can calculate $C_{mj}^{(-1)}$ from (24).

Let us take $\sigma_i = S_j^{-1}\delta_{ij} + S_m^{-1}\delta_{im}$, $1 \leq i \leq n$ in (24). Then

$$C_{mj}^{(-1)} \leq -\frac{C_{jj}^{(-1)} + C_{mm}^{(-1)}}{2} + \frac{\tilde{C}_{jj}^{(-1)} + 2\tilde{C}_{jm}^{(-1)} + \tilde{C}_{mm}^{(-1)}}{2} \tag{25}$$

where $C_{jm}^{(-1)}$ is defined in (13). If lower bounds on $C_{jj}^{(-1)}$ are known one can obtain upper bounds on $C_{mj}^{(-1)}$ from (25).

Chapter 4. Numerical Examples

§1. Introduction

Algorithms for calculating electrostatic fields, or linear functionals
of these fields such as electrical capacitances, given in Chapters 2 and
3, are reduced to calculations of certain multiple integrals. From the
point of view of numerical analysis one should integrate functions with at
worst weak singularities. The numerical integration of such functions is
a problem of independent interest. It has been discussed in detail for
functions of one variable [38], but not much is known about multidimensional integrals of functions with weak singularities. The basic idea in
the one-dimensional case is to integrate explicitly the singular part of
the integrand and thus to reduce the problem to the integration of a smooth
function. This problem is well understood.

In the multidimensional case it seems that the first step in the above
program was not discussed. In this chapter two problems of practical interest will be solved. First the table of the capacitances of the circular metallic cylinder will be given. Secondly the table of the capacitances
of the metallic parallelepiped of arbitrary shape will be given. Both results seem to be new. Special cases such as the capacitance of a cube,
disk, or very long cylinder will be compared with previously published results. It seems that the numerical results show that the formulas for
calculating the capacitance given in Chapter 3 are particularly efficient.

§2. Capacitance of a Circular Cylinder

Let $2L$ be the length and a be the radius of a metallic cylinder.
Let $C_1 = C/(2L)$ and $\ell = La^{-1}$. The capacitance per unit length C_1 is
given in Fig. 1, and Fig. 2 as a function of ℓ, $0.1 \le \ell \le 10$. The

42

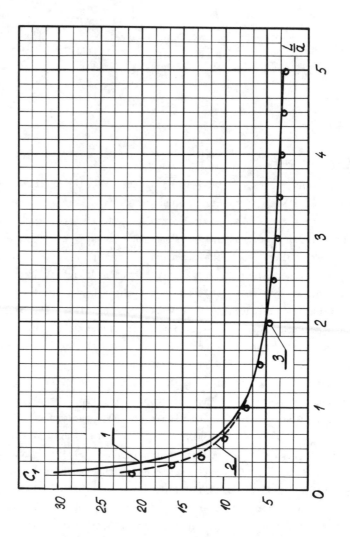

Figure 1

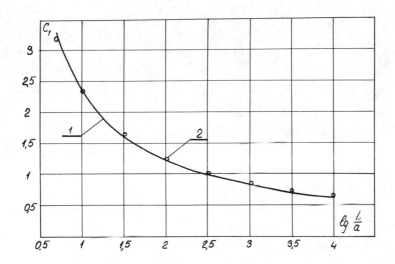

Figure 2

capacitance C was calculated using formula (3.1.12) with n = 0 and n = 1. It turned out that for $\ell \geq 5$, n = 0 this formula gives a value which agrees within 1% with the capacitance of a hollow metallic tube with the same geometry. Numerical calculation of the capacitance of such a tube was given in [10]. For $1 \leq \ell \leq 5$, n = 0 the difference (i.e., the relative error) is at most 3%. For $\ell \geq 1$ and n = 1 the difference is at most 1%, while for $0.1 \leq \ell \leq 1$, n = 1 the difference is at most 3%. For $\ell \leq 0.1$ the asymptotic formula holds

$$C_1 = 4\varepsilon_e \ell^{-1} \tag{1}$$

with the relative error at most 3%. This formula follows from the known formula $C = 8a\varepsilon_e$ for the capacitance of the metallic disk of radius a and the definition $C_1 = C/(2L)$. As $\ell \to 0$ the accuracy of formula (1) increases. For $\ell \geq 10$ the formula

$$C_1 = 4\pi\varepsilon_e (\Omega^{-1} + 0.71\Omega^{-3}), \qquad \Omega \equiv 2[\ln(4\ell) - 1] \tag{2}$$

holds [10] with error at most 1%. For $\ell \geq 4$ formula (2) holds with the error at most 3.5%. For $0.1 \leq \ell \leq 4$ the formula

$$C_1 = \frac{2\pi^2 \varepsilon_e}{\ln(16\ell^{-1})} \tag{3}$$

holds with error at most 3.5%. Thus formulas (1)-(3) give C_1 for any ℓ with the error at most 3.5%. An unexpected observation is that

$$\frac{C_{1tube}}{C_{1cylinder}} = \frac{\pi^2}{2\ln(16\ell^{-1})} = \frac{4.93}{\ln(16\ell^{-1})}, \qquad \ell \ll 1. \tag{4}$$

This formula follows from (1) and (3). Formula (3) is the asymptotic formula for the capacitance of the tube for $\ell \ll 1$. For $\ell = 0.1$ the ratio (4) is equal to 0.98. This ratio is equal to 0.5 for $\ell^{-1} = 1250$. Thus the capacitance per unit length of the metallic cylinder is nearly equal to that of the tube for $\ell \geq 0.1$.

§3. Capacitance of a Parallelepiped of Arbitrary Shape

Let a parallelepiped have edges

$$A_1 \leq A_2 \leq A_3, \tag{1}$$

let V denote its volume, set

$$\lambda = V^{1/3} = (A_1 A_2 A_3)^{1/3}, \tag{2}$$

and let $C_\lambda = C(A_1, A_2, A_3)$ be its capacitance. Let

$$a_j = A_j \lambda^{-1}, \quad 1 \leq j \leq 3; \quad a_1 \leq a_2 \leq a_3, \quad a_1 a_2 a_3 = 1. \tag{3}$$

It is clear that

$$C_\lambda = \lambda \cdot C, \tag{4}$$

where C is the capacitance of the parallelepiped with sides a_1, a_2, a_3 and unit volume.

Therefore it is sufficient to tabulate $C(a_1, a_2, a_3)$, where a_j, $1 \leq j \leq 3$ satisfy (3).

Some long but not complicated calculations using formula (3.1.12) with $n = 0$ lead to the formula

$$\frac{C_\lambda}{4\pi\varepsilon_0} \approx \frac{S^2}{J} \, , \tag{5}$$

where

$$S = 2(A_1 A_2 + A_1 A_3 + A_2 A_3) \tag{6}$$

and

$$
\begin{aligned}
J = & \frac{4}{3} \sum_{i=1}^{3} \left[d\left(D^2 - \frac{S}{2} - \frac{3V}{A_i} \right) - A_i^3 \right] \ln \frac{D - A_i}{D + A_i} \\
& + \frac{4}{3} \sum_{i=1}^{3} \sum_{j \neq i} \frac{V^2}{A_i^2 A_j} \left(3 + \frac{V}{A_i A_j^2} \right) \ln \frac{\sqrt{D^2 - A_i^2} + A_j}{\sqrt{D^2 - A_i^2} - A_j} \\
& - \frac{8}{3} \sum_{i=1}^{3} \left(D^2 - A_i^2 - \frac{2V}{A_i} \right) \sqrt{D^2 - A_i^2} \\
& - \frac{8}{3} SD + \frac{16}{3} \left[d(D^2 - \frac{S}{2}) + 3V \right] - \frac{8}{3} \sum_{i=1}^{3} A_i (A_i^2 + 3S) \arctan \frac{V}{A_i^2 D} \, ,
\end{aligned} \tag{7}
$$

where

$$D = \left(\sum_{i=1}^{3} A_i^2 \right)^{1/2}; \quad d = \sum_{i=1}^{3} A_i; \quad S = 2V \sum_{i=1}^{3} \frac{1}{A_i}; \quad V = \prod_{i=1}^{3} A_i.$$

Let us describe a way to tabulate

$$\tilde{C} \equiv \frac{C}{4\pi\varepsilon_e} \tag{8}$$

It follows from (3) that

$$0 \leq a_1 \leq 1. \tag{9}$$

Let

$$a_1 = kn^{-1}, \quad 1 \leq k \leq n \tag{10}$$

where n is an integer which defines the table. Let

$$a_2 = jn^{-1}, \quad j \geq k. \tag{11}$$

Then

$$a_3 = \frac{1}{a_1 a_2} = \frac{n^2}{kj}, \quad k \leq j. \tag{12}$$

From (3) it follows that $jn^{-1} \leq n^2 (kj)^{-1}$. Thus

$$\frac{k}{n} \leq \frac{j}{n} \leq \sqrt{n/k}. \tag{13}$$

Therefore

$$a_1 \leq a_2 \leq \frac{1}{\sqrt{a_1}} \, . \tag{14}$$

For fixed a_1 and a_2, the parameter a_3 is uniquely determined by (12). This means that $\tilde{C}$ can be tabulated as a function of a_1 and a_2. In Table 1 the results are given for $n = 10$. In the horizontal line the values of a_1 are given. In the vertical line the values of a_2 are given. At the intersections the values of $\tilde{C}(a_1, a_2)$ are given. If zero stands at the intersection, this means that for the given a_1 the chosen a_2 is not allowed by (14).

Let us formulate an algorithm for calculating C_λ for an arbitrary parallelepiped.

Step 1. Order the sides of the parallelepiped as shown in (1) and calculate λ from (2) and a_1 and a_2 from (3).

Step 2. Find the numbers closest to a_1 and a_2 in the horizontal and vertical line of Table 1 respectively. Find $C(a_1, a_2)$ in this table.

Step 3. Find C_λ from (4) and (8).

Example 1. Let $A_1 = 1$, $A_2 = 2$, $A_3 = 4$. Then $V = 8$, $\lambda = 2$, $a_1 = 0.5$, $a_2 = 1$, $\tilde{C} = 0.70633$. Thus $C_\lambda = 8\pi\varepsilon_e \cdot 0.70633 \approx 17.7514\varepsilon_e$.

Example 2. Let $A_1 = A_2 = A_3 = 1$, i.e., we have a unit cube, $a_1 = a_2 = a_3 = 1$, $V = 1$, $\lambda = 1$. From Table 1 one find $C = 4\pi\varepsilon_e \cdot 0.649$.

References [10] and [27] mention about 17 papers dealt with the test problem of calculating the capacitance of a cube. The best results reported in [27] and obtained by means of some complicated calculations with harmonic polynomials with the symmetry group of a cube, were that the capacitance C of the unit cube satisfies

$$0.632 < \frac{C}{4\pi\varepsilon_e} < 0.710, \qquad \frac{C}{4\pi\varepsilon_e} \approx 0.646. \tag{15}$$

From (3.1.12) and (3.1.15) it follows that the value $C/(4\pi\varepsilon_e) = 0.649$ is not only an approximation to $C/(4\pi\varepsilon_e)$ but also a lower bound. One can see that for a cube formula (3.1.12) gave a good result even for $n = 0$.

Example 3. Let $A_1 = 0$, $A_2 = 2$, $A_3 = 5$. This is the case of a thin rectangular metallic plate. Since the smallest $a_1 = 0.1$ in Table 1, we take $A_1 = 0.1$, $A_2 = 2$, $A_3 = 5$ and find $C = 4\pi\varepsilon_e \cdot 1.18577$. This

TABLE 1. Table of the Capacitances $\tilde{C} = C/(4\pi\epsilon_e)$ of the Unit Parallelepiped

a_2	a_1									
	0.1	0.2	0.3	0.4	0.5	0.6	0.7	0.8	0.9	1.0
0.1000	7.00313									
0.2000	4.12588	2.47336								
0.3000	3.08985	1.88108	1.44955							
0.4000	2.54667	1.57289	1.22690	1.04998						
0.5000	2.21009	1.38371	1.09186	0.94404	0.85669					
0.6000	1.98066	1.25629	1.00224	0.87489	0.80064	0.75381				
0.7000	1.81434	1.16528	0.93938	0.82736	0.76294	0.72320	0.69733			
0.8000	1.68855	1.09767	0.89367	0.79366	0.73693	0.70237	0.68067	0.66708		
0.9000	1.59040	1.04604	0.85968	0.76936	0.71883	0.68859	0.67007	0.65894	0.65278	
1.0000	1.51203	1.00586	0.83405	0.75174	0.70633	0.67963	0.66373	0.65463	0.65011	0.6488
1.1000	1.44832	0.97417	0.81461	0.73906	0.69794	0.67461	0.66050	0.65312		
1.2000	1.39582	0.94897	0.79989	0.73010	0.69264	0.67145				
1.3000	1.35207	0.92885	0.78885	0.72404	0.68974					
1.4000	1.31531	0.91277	0.78074	0.72029	0.68872					
1.5000	1.28423	0.89998	0.77499	0.71837						
1.6000	1.25784	0.88990	0.77118							
1.7000	1.23534	0.88206	0.76896							
1.8000	1.21614	0.87611	0.76808							
1.9000	1.19975	0.87173								
2.0000	1.18577	0.86878								
2.1000	1.17387	0.86698								
2.2000	1.16380	0.96620								
2.3000	1.15532									
2.4000	1.14825									
2.5000	1.14243									
2.6000	1.13771									
2.70000	1.13399									
2.8000	1.13115									
2.9000	1.12911									
3.0000	1.12780									
3.1000	1.12714									

agrees with the value given in [6].

Example 4. Consider the square thin plate: $A_1 = 0.1$, $A_2 = A_3 = 1$. Let $a_1 = 0.1$, $a_2 = a_3 = 3.16$. Then $a_1 a_2 a_3 = 1$ and from Table 1 one finds $C/(4\pi\varepsilon_e) = 1.12714$. For the capacitance of the thin plate with the unit side one finds $C^{(1)}/(4\pi\varepsilon_e) = 1.12714/3.16 = 0.3566$. This agrees with the value 0.360 given in [10].

Remark 1. Table 1 shows that among all parallelepiped with the fixed volume the cube has the minimal capacitance. This can be proved, but the proof [21] is not elementary. The error in the calculation of the capacitances in Table 1 is at most 2%.

§4. Interaction Between Conductors

Let two conducting balls of radius a be charged up to potential V. Then $Q = C_{11}V + C_{12}V$, $Q = C_{21}V + C_{22}V$ and by symmetry $C_{11} = C_{22}$, $C_{12} = C_{21}$. Let us join these balls. The electrostatic equilibrium will be preserved since the potentials of the balls are the same. Let $\tilde{C}$ denote the capacitance of the joined balls. Then $\tilde{C} = 2Q/V = 2(C_{11} + C_{12})$. Let C be the capacitance of a single ball. Then $\tilde{C}/(2C) = C_{11} + C_{12}/C$. Let d be the distance between the centers of the balls. Then the numerical results [10] give $\tilde{C}/(2C) = 0.75$ if $2ad^{-1} = 0.5$; $\tilde{C}/(2C) = 0.91$ if $2ad^{-1} = 0.2$; $\tilde{C}/(2C) = 0.71$ if $2ad^{-1} = 0.9$. Therefore one makes the error of at most 25% if one neglects the interaction of the conductors if $a \leq 0.25d$ and one makes the error of at most 10% if $a \leq 0.1d$.

Chapter 5. Calculating the Polarizability Tensor

§1. Calculating the Polarizability Tensor of a Solid Body

1. If a solid conductor is placed in an exterior homogeneous electrostatic field E, then on its surface the induced charge distribution
$\sigma(t)$ appears. Therefore the conductor acquires the dipole moment

$$P_i = \int_\Gamma t_i \sigma(t) dt, \tag{1}$$

where t_i is the ith coordinate of the radius vector t of the point t
at the surface Γ of the conductor. Since the equations of electrostatics are linear, there is a linear relation between P and E:

$$P_i = \alpha_{ij} \epsilon_e V E_j, \tag{2}$$

with summation over the repeated index, where V is the volume of the
conductor, ϵ_e is the dielectric permittivity of the exterior medium, the
matrix α_{ij} is called the polarizability tensor. The dipole moment is
interesting in many applications, especially in scattering theory (see
Chapter 7).

A more general definition of the dipole moment is as follows. Let
$\phi_0 = -(E,x)$ be the potential of the exterior homogeneous field, $\phi = \phi_0 + u$ be the potential of the total field. If the obstacle is finite,
then

$$u \sim \frac{(P,x)}{4\pi\epsilon_e |x|^3} \quad \text{as} \quad |x| \to \infty. \tag{3}$$

We assume here that the obstacle is electroneutral. The vector P is
called the dipole moment induced on the obstacle by the exterior field E.

2. Let the obstacle be a homogeneous body with dielectric constant ε. Put

$$\gamma = \frac{\varepsilon - \varepsilon_e}{\varepsilon + \varepsilon_e} . \tag{4}$$

The polarizability tensor is defined by the formula

$$P_i = \alpha_{ij}(\gamma)\varepsilon_e V E_j . \tag{5}$$

If $\varepsilon = \infty$ then $\gamma = 1$, $\alpha_{ij}(1) = \alpha_{ij}$ where α_{ij} is the polarizability tensor of the perfect conductor with the same shape. If $\varepsilon = 0$, then $\gamma = -1$, $\alpha_{ij}(-1) \equiv \beta_{ij}$, where β_{ij} is the magnetic polarizability tensor (the polarizability tensor of the insulator). Our aim is to give approximate analytical formulas for calculating $\alpha_{ij}(\gamma)$. Let us introduce some notations. Let

$$b_{ij}^{(0)} = V\delta_{ij}, \quad \delta_{ij} = \begin{cases} 1, & i = j, \\ 0, & i \neq j, \end{cases} \tag{6}$$

$$b_{ij}^{(1)} = \int_{\Gamma}\int_{\Gamma} \frac{N_i(t)N_j(s)}{r_{st}} \, ds dt, \tag{7}$$

where $N_i(t)$ is the ith component of the outer unit normal to Γ at the point t,

$$b_{ij}^{(m)} = \int_{\Gamma}\int_{\Gamma} ds dt \, N_i(t)N_j(s) \underbrace{\int_{\Gamma}\cdots\int_{\Gamma}}_{m-1} \frac{1}{r_{st_{m-1}}} \psi(t_1,t)\psi(t_2,t_1)\cdots$$
$$\psi(t_{m-1},t_{m-2})dt_1\cdots dt_{m-1}, \tag{8}$$

where

$$\psi(t,s) = \frac{\partial}{\partial N_t} \frac{1}{r_{st}} .$$

Let

$$\alpha_{ij}^{(n)}(\gamma) = \frac{2}{V} \sum_{m=0}^{n} \frac{(-1)^m}{(2\pi)^m} \frac{\gamma^{n+2} - \gamma^{m+1}}{\gamma - 1} b_{ij}^{(m)}, \quad n > 0. \tag{9}$$

In particular

$$\alpha_{ij}^{(1)}(\gamma) = 2(\gamma + \gamma^2)\delta_{ij} - \frac{\gamma^2}{\pi V} b_{ij}^{(1)}, \tag{10}$$

$$\alpha_{ij}^{(1)} = 4\delta_{ij} - \frac{1}{\pi V} b_{ij}^{(1)}, \tag{11}$$

$$\beta_{ij}^{(1)} = - \frac{1}{\pi V} b_{ij}^{(1)} . \tag{12}$$

Note that $b_{ij}^{(m)}$ depends only on the geometry of the body.

Theorem 1. *The following estimate holds*

$$|\alpha_{ij}(\gamma) - \alpha_{ij}^{(n)}(\gamma)| \leq cq^n, \quad 0 < q < 1, \quad -1 \leq \gamma \leq 1, \tag{13}$$

where $c > 0$ *and* q *are constants which depend only on the shape of* Γ *and on* γ.

Remark 1. From (9) for $\varepsilon = \infty$ (i.e., $\gamma = 1$) it follows that

$$\alpha_{ij}^{(n)} = \frac{2}{V} \sum_{m=0}^{n} \frac{(-1)^m}{(2\pi)^m} (n+1-m) b_{ij}^{(m)}, \tag{14}$$

and for $\varepsilon = 0$ (i.e., $\gamma = -1$) it follows that

$$\beta_{ij}^{(n)} = \frac{1}{V} \sum_{m=0}^{n} \frac{(-1)^{n+m-1} - 1}{(2\pi)^m} b_{ij}^{(m)}. \tag{15}$$

Proof of Theorem 1. Let us define

$$P_i^{(n)} = \int_{\Gamma} t_i \sigma_n dt \tag{16}$$

where σ_n is defined in (2.2.1) with $\sigma_0 = -2\gamma\varepsilon_e (\partial\phi_0/\partial N)$,

$$|\sigma_n - \sigma| \leq cq^n, \quad 0 < q < 1 \tag{17}$$

where $c > 0$ and q depend on Γ and γ. From (2.2.1) it follows that

$$\sigma_n = \sum_{m=0}^{n} (-1)^m \gamma^m A^m (2\gamma(E,N))\varepsilon_e. \tag{18}$$

From (16) and (18) one obtains

$$P_i^{(n)} = \frac{2}{V} \sum_{m=0}^{n} \frac{(-1)^m \gamma^{m+1}}{(2\pi)^m} \int_{\Gamma} t_j B^m (N_j) dt \, V\varepsilon_e E_j, \tag{19}$$

where

$$B \equiv 2\pi A. \tag{20}$$

Therefore

$$\alpha_{ij}^{(n)}(\gamma) = \frac{2}{V} \sum_{m=0}^{n} \frac{(-1)^m \gamma^{m+1}}{(2\pi)^m} J_{ij}^{(m)}, \tag{21}$$

where

$$J_{ij}^{(m)} = \int_{\Gamma} t_i B^m (N_j) dt. \tag{22}$$

Let us prove that

$$J_{ij}^{(m)} = b_{ij}^{(m)} - 2\pi J_{ij}^{(m-1)}, \tag{23}$$

where $b_{ij}^{(m)}$ is defined in (8). We have

$$J_{ij}^{(0)} = \int_{\Gamma} t_i N_j(t) dt = \int_D \frac{\partial x_i}{\partial x_j} dx = V\delta_{ij} = b_{ij}^{(0)}, \tag{24}$$

and

$$\begin{aligned}
J_{ij}^{(1)} &= \int_{\Gamma} s_i \, B(N_j) ds - \int_{\Gamma} dt N_j(t) \int_{\Gamma} s_i \frac{\partial}{\partial N_s} \frac{1}{r_{st}} ds \\
&= \int_{\Gamma} dt N_j(t) \left(\int_{\Gamma} \frac{\partial s_i}{\partial N_s} \frac{ds}{r_{st}} - 2\pi t_i \right) = \int_{\Gamma}\int_{\Gamma} \frac{N_i(s)N_j(t)}{r_{st}} dsdt - 2\pi V\delta_{ij} \\
&= b_{ij}^{(1)} - 2\pi J_{ij}^{(0)}.
\end{aligned} \tag{25}$$

In a similar manner, one obtaines

$$J_{ij}^{(m)} = \int_{\Gamma} ds \, s_i B^m(N_j) = \int_{\Gamma} dt \, N_j(t) \int_{\Gamma} dt_1 \, \psi(t_1,t) \cdots \int_{\Gamma} dt_{m-1}\psi(t_{m-1},t_{m-2})$$
$$\left[\int_{\Gamma} \frac{N_i(s)ds}{r_{st_{m-1}}} - 2\pi(t_{m-1})_i \right] = b_{ij}^{(m)} - 2\pi J_{ij}^{(m-1)}. \tag{26}$$

From (26) it follows that

$$J_{ij}^{(m)} = \sum_{k=0}^{m} b_{ij}^{(k)} (2\pi)^{m-k}(-1)^{m-k} \tag{27}$$

Using (27) and (21) one finds that

$$\begin{aligned}
\alpha_{ij}^{(n)}(\gamma) &= \frac{2}{V} \sum_{m=0}^{n} \frac{(-1)^m \gamma^{m+1}}{(2\pi)^m} \sum_{k=0}^{m} b_{ij}^{(k)} (2\pi)^{m-k}(-1)^{m-k} \\
&= \frac{2}{V} \sum_{k=0}^{n} b_{ij}^{(k)} \frac{(-1)^k}{(2\pi)^k} \frac{\gamma^{n+2} - \gamma^{k+1}}{\gamma - 1}.
\end{aligned} \tag{28}$$

Estimate (13) follows from (17). Theorem 1 is proved. □

§2. The Polarizability Tensor of a Thin Metallic Screen

Let F be a thin metallic screen. Its polarizability tensor is de-fined as

$$P_i = \alpha_{ij} E_j \varepsilon_e, \qquad P_i = \int_F t_i \sigma(t) dt, \tag{1}$$

where $\sigma(t)$ is the distribution of the charge induced by the exterior homogeneous electrostatic field E. Let e_i, $1 \leq i \leq 3$ be the orthonormal unit vectors of the coordinate system, let $E = e_j$, and let $\phi_0 = -x_j$ be the potential corresponding to E. Then

$$P_i = \alpha_{ij}\epsilon_e. \tag{2}$$

Let $\sigma_n(t)$ be the approximate charge distribution constructed in (2.4.2). Then

$$P_i^{(n)} = \int_\Gamma t_i \sigma_n(t)dt \equiv \alpha_{ij}^{(n)}\epsilon_e. \tag{3}$$

Thus

$$\alpha_{ij}^{(n)} = \epsilon_e^{-1} \int_\Gamma t_i \sigma_i(t)dt. \tag{4}$$

Note that the index j is actually present in the right-hand side of (4) because $\sigma_n(t)$ is constructed for the initial field $E = e_j$, or for the initial potential $\phi_0 = -x_j$. Thus the calculation of the polarizability tensor is reduced to finding σ_n according to Theorem 2.4.3 and to the calculation of the six integrals in (4), $1 \leq i \leq j \leq 3$. The number of the integrals is six (and not nine) because $\alpha_{ij}^{(n)} = \alpha_{ji}^{(n)}$.

Consider the case in which F is a plane plate. Let e_3 be orthogonal to F. Then $\alpha_{i3} = \alpha_{3i} = 0$ and the polarizability tensor is defined by the three numbers α_{11}, α_{22}, and $\alpha_{12} = \alpha_{21}$.

§3. The Polarizability Tensors of a Flaky-Homogeneous Body or a System of Bodies

1. The integral equation for the surface charge densities induced by the initial field is given in Theorem 2.2.1. The nth approximation for the polarizability tensor of the flaky-homogeneous body is very complicated. Therefore only the first approximation will be considered. Let A_{ij} be the polarizability tensor

$$P_i = A_{ij}E_j\epsilon_e. \tag{1}$$

There is no factor V in this definition of A_{ij} because if the body is nonhomogeneous the matrix $\alpha_{ij} = A_{ij}V^{-1}$ does not depend only on the geometry of the body. For the dipole moment of the flaky-homogeneous body one has the formula

$$P_i = \sum_{j=1}^p \int_{\Gamma_j} t_i \sigma_j(t)dt. \tag{2}$$

Substituting $\sigma_j^{(n)}$ from Theorem 2.2.1 in (2) in place of σ_j yields the nth approximation to P_i,

$$P_i^{(n)} = \sum_{j=1}^{P} \int_{\Gamma_j} t_i \sigma_j^{(n)}(t)dt \equiv A_{ij}^{(n)} E_j \varepsilon_e. \tag{3}$$

Let us take $n = 1$. From (3) and Theorem 2.2.1, it follows that $(E = -\nabla \phi_0)$

$$P_i^{(1)} = \sum_{j=1}^{P} \varepsilon_e \int_{\Gamma_j} t_i \Big\{ [2\gamma_j N_q(t)E_q - 2\gamma_j^2 A_j(E_q N_q(t))]$$

$$- 2\gamma_j \gamma_m \sum_{m \neq j, m=1}^{P} T_{jm}(E_q N_q(t)) \Big\} dt$$

$$= \Big\{ \sum_{j=1}^{P} \alpha_{iq}^{(1)}(\gamma_j)V_j + \sum_{j=1}^{P} \sum_{m \neq j, m=1}^{P} \alpha_{iq}^{(j,m)} \Big\} E_q \varepsilon_e, \tag{4}$$

where V_j is the volume of the body inside Γ_j,

$$\alpha_{ij}^{(1)}(\gamma_j) = 2\delta_{ij}(\gamma_j + \gamma_j^2) - \frac{\gamma_j^2}{\pi V_j} b_{ij}^{(1)}, \tag{5}$$

γ_j is given in (1.3.35) (compare (5) and (1.10)), and

$$\alpha_{iq}^{(j,m)} = \begin{cases} -\dfrac{\gamma_j \gamma_m}{\pi} b_{ij}^{(j,m)} & , \quad j > m \\[2ex] -\dfrac{\gamma_j \gamma_m}{\pi} b_{iq}^{(j,m)} + 4\gamma_j \gamma_m V_m \delta_{iq}, & j < m, \end{cases} \tag{6}$$

where

$$b_{iq}^{(j,m)} = \int_{\Gamma_j} \int_{\Gamma_m} \frac{N_i(t)N_q(s)}{r_{st}} \, ds \, dt. \tag{7}$$

These formulas and their proof are quite similar to formulas (1.8)-(1.12). Therefore the proof is left to the reader. From (3)-(7) one finds

$$A_{iq}^{(1)} = \sum_{j=1}^{P} \alpha_{iq}^{(1)}(\gamma_j)V_j + \sum_{j=1}^{P} \sum_{m \neq j, m=1}^{P} \alpha_{iq}^{(j,m)}, \tag{8}$$

where $\alpha_{iq}^{(1)}(\gamma_j)$ and $\alpha_{iq}^{(j,m)}$ are defined in (5) and (6) respectively.

2. Let us derive an approximate formula for the polarizability tensor of a system of bodies. We use Theorem 2.2.2 in the same manner as Theorem 2.2.1 was used. Let us define the polarizability tensor of a system of bodies by

$$P_i = B_{ij} E_j \varepsilon_e. \tag{9}$$

Then, following the line of argument given in Sec. 1, one finds

$$B_{iq}^{(1)} = \sum_{j=1}^{n} \alpha_{iq}^{(1)}(\kappa_j) V_j + \sum_{j=1}^{p} \sum_{m \neq j, m=1}^{p} \tilde{\alpha}_{iq}^{(j,m)}, \tag{10}$$

where κ_j is defined in (1.3.32), $\alpha_{iq}^{(1)}(\kappa_j)$ is defined in (5) where κ_j should replace γ_j,

$$\tilde{\alpha}_{iq}^{(j,m)} = - \frac{\kappa_j \kappa_m}{\pi} b_{iq}^{(j,m)} \tag{11}$$

and $b_{iq}^{(j,m)}$ is defined in (7).

If the jth body is a perfect conductor then $\kappa_j = 1$.

§4. Variational Principles for Polarizability Tensors

1. The purpose of this section is to give variational principles for polarizability tensors and to show how some two-sided estimates for the polarizability tensors can be obtained from these principles.

Let $E = e_j$, where e_j is the coordinate unit vector, $\phi_0 = -x_j$, $E = -\nabla\phi_0$. Suppose that the body is a perfect conductor. Then the induced surface charge distribution $\sigma_j(t)$ satisfies the equation

$$\int_\Gamma \frac{\sigma_j(t) dt}{4\pi\varepsilon_e r_{st}} = U_j + s_j, \quad U_j = const, \tag{1}$$

and the electroneutrality condition

$$\int_\Gamma \sigma_j \, dt = 0. \tag{2}$$

The quantity U_j is the potential of the conductor. The induced dipole moment of the conductor is

$$P_i = \alpha_{iq} \varepsilon_e V E_q = \alpha_{ij} \varepsilon_e V = \int_\Gamma t_i \sigma_j(t) dt, \tag{3}$$

because $E_q = \delta_{jq}$. Therefore

$$V\alpha_{ij} = \varepsilon_e^{-1} \int_\Gamma t_i \sigma_j(t) dt, \quad \alpha_{ij} = \alpha_{ji}. \tag{4}$$

Note that (2) and (1) imply

$$\int_\Gamma U_j \sigma_j \, dt = 0. \tag{5}$$

From (1), (5), and (3.2.8) it follows that

$$V\alpha_{ij} = st \ 4\pi \ \frac{\int_\Gamma t_i\phi_j dt \ \int_\Gamma t_j\phi_i dt}{\int_\Gamma\int_\Gamma \dfrac{\phi_i(t)\phi_j(s)dsdt}{r_{st}}} \tag{6}$$

where the admissible functions satisfy (2). For $i = j$ the st in (6) can be replaced by max.

$$V\alpha_{jj} = \max \ 4\pi \left(\int_\Gamma t_j\phi_j dt\right)^2 \left(\int_\Gamma\int_\Gamma \frac{\phi_j(t)\phi_j(s)dsdt}{r_{st}}\right)^{-1}. \tag{7}$$

where again ϕ_j satisfies (2). Principle (7) allows one to find lower bounds for the diagonal elements of the polarizability tensor.

2. In order to find upper bounds for these elements we need another variational principle. The energy of the electrostatic field of the conductor is

$$U = \frac{\varepsilon_e}{2} \int_{D_e} |\nabla\phi_j|^2 dx \tag{8}$$

where ϕ_j is the secondary potential corresponding to the initial field $E = e_j$.

On the other hand the same energy is equal to

$$U = \frac{\varepsilon_e V}{2} \alpha_{jj}. \tag{9}$$

Indeed, $U = \frac{1}{2}(P,E) = \frac{\varepsilon_e V}{2} \alpha_{im}E_m E_i$ and since $E_m = \delta_{jm}$ one obtains (9). Thus

$$V\alpha_{jj} = \min \int_{D_e} |\nabla u|^2 dx, \tag{10}$$

where the admissible functions $u \in C^1(D_e)$ satisfy the condition

$$u|_\Gamma = U_j + s_j, \quad U_j = \text{const.} \tag{11}$$

The minimum in (10) is attained at the solution of the problem

$$\Delta\phi = 0 \ \text{ in } \ D_e, \quad u|_\Gamma = U_j + s_j, \quad \int_\Gamma \frac{\partial\phi}{\partial N}dt = 0, \quad \phi(\infty) = 0. \tag{12}$$

The variational principle (10)-(11) allows one to obtain upper bounds for α_{jj}.

Example. Let Γ be a sphere with radius a. By symmetry one concludes that $\alpha_{ij} = \alpha\delta_{ij}$, where $\alpha > 0$ is a scalar. Let $\phi_j(t) = Y_{ij}(t)$, where Y_{ij} are the spherical harmonics, $Y_{11} = \cos\theta$, $Y_{12} = \sin\theta\cos\phi$, $Y_{13} = \sin\theta\sin\phi$, and $t = (1,\theta,\phi)$. From (6) one finds that $\alpha_{ij} = 0$ for $i \neq j$. For $i = j$ it follows from (7) that $\alpha_{jj} = \alpha = 3$. We omit some standard calculations which lead to this result. In this example we obtained the exact value of α because of the symmetry.

3. Suppose that $V \to 0$ and the body tends to a thin screen F with the edge L. Then the variational principles (6), (7), and (10)-(11) remain valid but the admissible functions should satisfy the edge condition. The tensor

$$\lim_{V \to 0} \alpha_{ij} V = \tilde{\alpha}_{ij} \tag{13}$$

is the polarizability tensor of the screen F. Therefore the derivation of the variational principles for the electric polarizability tensor of the metallic screen has no new points.

4. Let us derive some variational principles for the magnetic polarizability tensor β_{ij}. This tensor is defined as follows.
Consider the boundary value problem

$$\Delta\phi = 0 \text{ in } D_e, \quad -\frac{\partial\phi}{\partial N_e} = -\frac{\partial t_j}{\partial N} = -N_j(t), \quad \phi(\infty) = 0. \tag{14}$$

This problem is a mathematical formulation of the physical problem of the magnetic field around a superconductor (i.e., a body D inside which the magnetic induction $B = 0$). On the surface Γ of this body $B_N|_\Gamma = 0$. Outside the body div $B = 0$, curl $H = 0$, $B = \mu_0 H$ in D_e where μ_0 is the magnetic permeability of the exterior medium. If $H = e_j - \nabla\phi = \nabla(x_j - \phi)$ then the condition $B_N|_\Gamma = 0$ can be written as

$$\frac{\partial(x_j-\phi)}{\partial N_e}\bigg|_\Gamma = 0, \quad \text{or} \quad -\frac{\partial\phi}{\partial N_e} = -N_j \quad \text{on } \Gamma, \tag{15}$$

which is the same condition as in (14). Let

$$\phi = \phi_j = \int_\Gamma \frac{\sigma_j(t)dt}{4\pi\mu_0 r_{xt}}. \tag{16}$$

Then from (15) it follows that

$$\sigma_j = A\sigma_j - 2\mu_0 N_j(t), \quad A\sigma = \int_\Gamma \frac{\partial}{\partial N_s}\frac{1}{2\pi r_{st}}\sigma(t)dt \tag{17}$$

and

$$\sigma_j = \left(\frac{\partial \phi_j}{\partial N_i} - \frac{\partial \phi_j}{\partial N_e}\right)\mu_0. \tag{18}$$

The magnetic polarizability tensor is defined by the equation

$$V\beta_{pj} = \mu_0^{-1} \int_\Gamma t_p \sigma_j(t)dt, \tag{19}$$

where V is the volume of the body D.

If we substitute (18) into (19), we obtain

$$V\beta_{pj} = \int_\Gamma t_p \left(\frac{\partial \phi_j}{\partial N_i} - \frac{\partial \phi_j}{\partial N_e}\right)dt = \int_\Gamma \frac{\partial t_p}{\partial N} \phi_j dt - \int_\Gamma t_p \frac{\partial t_j}{\partial N} dt$$

$$= \int_\Gamma \frac{\partial \phi_p}{\partial N_e} \phi_j dt - \delta_{pj}V = -\int_{D_e} \nabla\phi_p \nabla\phi_j dx - \delta_{pj}V. \tag{20}$$

In particular

$$V\beta_{jj} + V = -\int_{D_e} |\nabla\phi_j|^2 dx, \tag{21}$$

$$V\beta_{pj} = \int_\Gamma \frac{\partial t_p}{\partial N} \phi_j dt - V\delta_{pj}. \tag{22}$$

The operator $-\partial/\partial N_e$ is nonnegative definite on the set of functions on Γ which are restrictions on Γ of harmonic functions defined in D_e and vanishing at infinity. This follows from the Green formula

$$-\int_\Gamma v \frac{\partial u}{\partial N_e} dt = \int_{D_e} \nabla u \nabla v \, dx = -\int_\Gamma u \frac{\partial v}{\partial N_e} dt. \tag{23}$$

Therefore formulas (15) and (3.2.8) yield

$$-V\beta_{pj} = st \frac{\int_\Gamma N_p(t)u_j(t)dt \int_\Gamma N_j(t)u_p(t)dt}{-\int_\Gamma \frac{\partial u_p}{\partial N_e} u_j dt} + V\delta_{pj}, \tag{24}$$

where the admissible functions $u_j(t)$ are harmonic in D_e and $u_j(\infty) = 0$. If $p = j$ then st in (24) can be replaced by max, obtaining

$$-(V+V\beta_{jj}) = max\left\{\left(\int_\Gamma N_j(t)u_j(t)dt\right)^2 \left(-\int_\Gamma \frac{\partial u_j}{\partial N_e} u_j dt\right)^{-1}\right\}, \tag{25}$$

or

$$-(V+V\beta_{jj}) = max\left\{\left(\int_\Gamma N_j u_j dt\right)^2 \left(\int_{D_e} |\nabla u_j|^2 dx\right)^{-1}\right\}. \tag{26}$$

The maximum in (25), (26) is attained at the solution to (14).

Remark 1. Formulas (25), (26) remain valid if the admissible functions u are not necessarily harmonic in D_e but are arbitrary functions $u \in C^1(D_e)$, $u(\infty) = 0$.

Proof: From (21) and (26) it follows that (26) can be written as

$$\int_{D_e} |\nabla u|^2 dx \int_{D_e} |\nabla \phi_j|^2 dx \geq \left(\int_{D_e} \nabla \phi_j \nabla u_j dx\right)^2 = \left(\int_{\Gamma} N_j(t)u_j dt\right)^2. \qquad (27)$$

The equality in (27) follows from Green's formula.

Inequality (27) is just an instance of the Cauchy inequality and is valid for any u, ϕ_j such that $\nabla u \in L^2(D_e)$, $\nabla \phi_j \in L^2(D_e)$, not necessarily harmonic in D_e. □

Exercise. Prove that

$$-2\pi V\beta_{pj} = \delta_{pj} + st \frac{\int_\Gamma \int_\Gamma N_p(t)\sigma_j(s)\frac{dsdt}{r_{st}} \int_\Gamma \int_\Gamma N_j(t)\sigma_p(s)\frac{dsdt}{r_{st}}}{\int_\Gamma \int_\Gamma \left\{\sigma_p(t) - A\sigma_p(t)\right\}\frac{\sigma_j(s)dsdt}{r_{st}}},$$

where A is defined in (17) and the admissible functions $\sigma_j(t) \in C(\Gamma)$.

Remark 2. Principle (26) allows one to obtain upper bounds for β_{jj}.

In order to obtain some lower bounds for β_{jj} the variational principle

$$-V - V\beta_{jj} = \min \int_{D_e} |q_j|^2 dx, \qquad (28)$$

where q_j can be used, is an arbitrary vector field such that the integral (28) converges and

$$\text{div } q_j = 0 \quad \text{in } D_e; \quad . \quad (q_j,N) = N_j(t) \quad \text{on } \Gamma. \qquad (29)$$

Proof: In order to prove principle (28)-(29), note that

$$\int_{D_e} |q-\nabla \phi_j|^2 dx = \int_{D_e} |q|^2 dx + \int_{D_e} |\nabla \phi_j|^2 dx - 2\int_{D_e} q\nabla \phi_j dx \qquad (30)$$

and

$$\int_{D_e} q\nabla \phi_j dx = \int_{D_e} \text{div}(q\phi_j)dx - \int_{D_e} \phi_j \text{ div } qdx = -\int_{\Gamma} (q,N)\phi_j dt$$

$$= -\int_{\Gamma} N_j\phi_j dt = -\int_{\Gamma} \frac{\partial \phi_j}{\partial N_e} u_j dt = \int_{D_e} |\nabla \phi_j|^2 dx. \qquad (31)$$

From (30), (31), and (21) it follows that

$$\int_{D_e} |q - \nabla\phi_j|^2 dx = \int_{D_e} |q|^2 dx - \int_{D_e} |\nabla\phi_j|^2 dx = \int_{D_e} |q|^2 dx + V + V\beta_{jj} \quad (32)$$

provided that q satisfies (29). Principle (28) follows from (32). The minimum in (28) is attained at $q_j = \nabla\phi_j$, where u_j is the solution to (14). □

5. <u>Magnetic polarizability of screens</u>. In connection with magnetic polarizability, the screen is a model of a thin superconductor or a perfect magnetic film. The latter case is of interest because thin magnetic films are parts of the memory elements of computers and the state of magnetic polarization of the film is the state which is stored in the memory.

Let us denote the magnetic polarizability tensor of the screen by

$$\tilde{\beta}_{ij} = \lim_{V \to 0} V\beta_{ij}. \quad (33)$$

This definition is similar to (13). The new point, in comparison with Sec. 3, is that if Γ is an unclosed surface one cannot look for a solution to problem (14) of the form (16). Indeed the normal derivative of the potential of a single layer (16) has a jump when x crosses Γ, while the boundary condition in (14) shows that the normal derivative is continuous when x crosses Γ. Therefore let us look for the solution of (14), in the case when the body is an unclosed thin surface F, of the form

$$\psi_j = \int_F \eta_j(t) \frac{\partial}{\partial N_t} \frac{1}{4\pi\mu_0 r_{xt}} dt. \quad (34)$$

It is known [7] that $\partial\psi_j/\partial N$ is continuous when x crosses F provided that the surface F is smooth. We have

$$\psi_j \sim \frac{(M_j, x)}{4\pi\mu_0 |x|^3}, \quad |x| \to \infty, \quad (35)$$

where

$$M_j \equiv \int_F \eta_j(t) N(t) dt. \quad (36)$$

The vector M is the induced magnetic moment. In particular,

$$M_{jj} = \int_F \eta_j(t) N_j(t) dt. \quad (37)$$

Since the initial field H has the potential $\phi_0 = -x_j$, we have

$$M_{jj} = \mu_0 \tilde{\beta}_{jj} H_j = \mu_0 \tilde{\beta}_{jj}. \quad (38)$$

Thus

$$\tilde{\beta}_{jj} = \mu_0^{-1} \int_F \eta_j(t) N_j(t) dt \qquad (39)$$

$$\tilde{\beta}_{pj} = \tilde{\beta}_{jp} = \mu_0^{-1} \int_F \eta_j(t) N_p(t) dt. \qquad (40)$$

Let us consider the boundary condition

$$- \frac{\partial \psi_j}{\partial N} = -N_j \quad \text{on} \quad \Gamma \qquad (41)$$

as an equation for η_j. The function ψ_j must satisfy the edge condition which can be formulated for this problem as

$$\lim_{\rho \to 0} \int_{S_\rho} \psi_j \frac{\partial \psi_j}{\partial N} ds = 0, \qquad (42)$$

where S_ρ is the surface of the torus generated by a disk of radius ρ whose center moves along the edge L of F so that the disk is perpendicular to L. Condition (42) allows one to integrate over F as if F were a closed surface. Namely $\int_F = \int_{F_+} + \int_{F_-}$, where $F_+(F_-)$ is the exterior (interior) side of F. It does not matter which of the two sides is chosen as exterior. From (26) it follows when $V \to 0$ that

$$-\tilde{\beta}_{jj} = \max \left\{ \left(\int_F N_j(t) u_j(t) \, dt \right)^2 \left(\int_{D_e} |\nabla u_j|^2 dx \right)^{-1} \right\} \qquad (43)$$

where the maximum is taken over the set of harmonic functions satisfying the edge condition (42). For example, one can take the admissible functions of the form (34). The surface F is the surface of discontinuity for the admissible functions.

Passing to the limit $V \to 0$ in (28) yields

$$-\tilde{\beta}_{jj} = \min \int_{D_e} |q_j|^2 dx, \qquad (44)$$

where q_j satisfies (29) and the edge condition (1.1.18). Principles (43) and (44) allow one to obtain lower and upper bounds for $\tilde{\beta}_{jj}$ respectively.

From (10), (41), and (3.2.8) it follows that

$$-\beta_{pj} = \text{st} \frac{\int_F N_p(t) \eta_j(t) dt \int_F N_j(t) \eta_p(t) dt}{-\int_F \frac{\partial}{\partial N_t} \left\{ \int_F \eta_p(s) \frac{\partial}{\partial N_s} \frac{1}{4\pi r_{st}} ds \right\} \eta_j(t) dt}, \qquad (45)$$

where the admissible functions $\eta_j(t)$ should satisfy the edge condition (1.1.17). Principle (45) holds also for closed surfaces, in which case $\tilde{\beta}_{pj} = V\beta_{pj}$.

The integral in the denominator of (45) can be transformed by means of the identity [8]:

$$\frac{\partial}{\partial N_t} \int_F \eta(s) \frac{\partial}{\partial N_s} \frac{1}{r_{st}} \, ds = \int_F ([N_s, \hat{\nabla}_s \eta], \, [N_t, \hat{\nabla}_s r_{st}^{-1}]) \, ds, \tag{46}$$

where $\hat{\nabla}$ is the surface gradient and $[a,b]$ is the vector product.

6. <u>Polarizability tensors for plane screens</u>. Let the x_3-axis be perpendicular to the screen. If $\Gamma = F$ in (6), (7) and F is a plane domain on the (x_1, x_2) plane then $\tilde{\alpha}_{i3} = \tilde{\alpha}_{3i} = 0$, $1 \le i \le 3$.

Similarly from (40) and (43) it follows that only $\tilde{\beta} \equiv \tilde{\beta}_{33} \ne 0$ if $\Gamma = F$ is a plane screen. From (43) it follows that

$$-\tilde{\beta} = \max\left\{\left(\int_F u \, dt\right)^2 \left(\int_{D_e} |\nabla u|^2 dx\right)^{-1}\right\}, \tag{47}$$

where the admissible functions satisfy the edge condition, vanish at infinity, and are harmonic.

From (44) it follows that

$$-\tilde{\beta} = \min \int_{D_e} |q|^2 dx, \tag{48}$$

where the admissible vectors satisfy the conditions

$$\text{div } q = 0 \quad \text{in } D_e, \quad q_3|_F = 1. \tag{49}$$

<u>Exercise</u>. Derive from (24) that

$$-\tilde{\beta} = \max\left\{\left(\int_F u(t) \, dt\right)^2 \left(\int_F \int_F \frac{\hat{\nabla}_t u(t) \hat{\nabla}_s u(s) \, ds \, dt}{4\pi r_{st}}\right)^{-1}\right\} \tag{50}$$

7. The variational principles, i.e., principles involving a maximum or minimum, were derived only for the diagonal elements of the polarizability tensors. Nevertheless they allow one to obtain two-sided estimates for any elements of the tensors.

To do so one can use the transformation properties of tensors and take into account that any element of a selfadjoint matrix is a linear combination of its diagonal elements in the coordinates in which the matrix is diagonal.

Chapter 6. Iterative Methods of Solving Some Integral Equations Basic in the Theory of Static Fields: Mathematical Results

§1. Iterative Methods of Solving the Fredholm Equations of the Second Kind at a Characteristic Value

The aim of this chapter is to provide in abstract setting some results which justify the iterative processes given in Chapter 2.

1. Let A be a linear compact operator on a Hilbert space H, λ_n, ϕ_n its characteristic values and eigenelements, $\phi_n = \lambda_n A \phi_n$, $|\lambda_1| < |\lambda_2| \leq |\lambda_3| \leq \ldots$. Let $G_1 \equiv \{\psi: (I - \bar{\lambda}_1 A^*)\psi = 0\}$ and $G_1^{\perp}$ its orthogonal complement in H. The equation

$$g - \lambda_1 A g = f \tag{1}$$

is solvable if and only if $f \in G_1^{\perp}$.

Main assumption: λ_1 is semisimple. $\tag{2}$

This means that the pole $\lambda = \lambda_1$ of the resolvent $(I - \lambda A)^{-1}$ is simple. This also means that the root subspace of A corresponding to λ_1 coincides with the eigensubspace of A corresponding to λ_1. The root subspace is defined as follows. Let $\phi = \lambda_1 A \phi$. Consider the equations

$$\phi^{(j+1)} - \lambda_1 A \phi^{(j+1)} = \phi^{(j)}, \quad j \geq 0, \quad \phi^{(0)} = \phi. \tag{3}$$

Only a finite number r of these equations are solvable [10]. If (3) has no solution for $j = 0$ then λ_1 is semisimple. If (3) is solvable for $0 \leq j \leq r$ and is not solvable for $j = r + 1$ then the set $\{\phi, \phi^{(1)}, \ldots, \phi^{(r)}\}$ is called the Jordan chain of length $r + 1$ associated with the pair (λ_1, ϕ). The elements $\phi^{(1)}, \ldots, \phi^{(r)}$ are called root vectors of A corresponding to λ_1. The linear span of all eigenvectors and root vectors corresponding to λ_1 is called the root space corresponding to λ_1. The linear span of eigenvectors corresponding to λ_1 is called the eigenspace corresponding to λ_1.

If the root space is one-dimensional then λ_1 is called simple. If the root space coincides with the eigenspace but has dimension greater than one then λ_1 is called semisimple. It can be proved that λ_1 is semisimple if and only if λ_1 is a simple pole of $(I - \lambda A)^{-1}$ [12]. It can also be proved that λ_1 is semisimple iff

$$(I - \lambda_1 A)^2 \phi = 0 \Rightarrow (I - \lambda_1 A)\phi = 0. \tag{4}$$

Lemma 1. If λ_1 is semisimple then equation (1) has at most one solution in $G_1^\perp$.

Proof: It is sufficient to prove that the homogeneous equation (1) has only trivial solutions in $G_1^\perp$. Suppose $\phi = \lambda_1 A \phi$, $\phi \in G_1^\perp$, $\phi \neq 0$. Since $G_1^\perp = R(I - \lambda_1 A)$, where $R(A)$ denotes the range of A, and since $G_1^\perp$ is closed, because A is compact, $\phi \in G_1^\perp$ implies there exists an f such that $\phi = (I - \lambda_1 A)f$. Therefore $(I - \lambda_1 A)^2 f = 0$ and from (4) it follows that $(I - \lambda_1 A)f = 0$, i.e., $\phi = 0$.

Remark 1. Equation (1) with semisimple λ_1 is important because most of the basic equations of electrostatics, magnetostatics, elastostatics, and hydrodynamics of ideal incompressible fluids are of this type. In practice f in (1) belongs to $G_1^\perp$ so that (1) is solvable. On the other hand, λ_1 is a characteristic value so that the resolvent $(I - \lambda A)^{-1}$ does not exist at $\lambda = \lambda_1$. Therefore equation (1) is an ill-posed problem: small perturbations of f can produce large perturbations in the solution or make equation (1) unsolvable. The theorems below show how to handle the situation.

Let $\{\phi_j\}$ be an orthonormal basis of $N(I - \lambda_1 A) \equiv \{\phi : (I - \lambda_1 A)\phi = 0\}$ and let $\{\psi_j\}$ be an orthonormal basis of $G_1 = N(I - \overline{\lambda}_1 A^*)$, $1 \leq j \leq m$. Let P be the orthogonal projection of H onto G. Define

$$B_\gamma g \equiv Ag + \gamma \sum_{j=1}^{m} (g, \psi_j)\psi_j \tag{5}$$

and

$$r_\gamma = \min(|\lambda_2|, |\lambda_1 (1 + \gamma \lambda_1)^{-1}|), \tag{6}$$

where γ is an arbitrary number which will be so chosen that $r_\gamma = |\lambda_2|$ e.g., $\gamma = -\lambda_1^{-1}$, and $(.,.)$ denotes the inner product in H.

Consider the equation

$$g = \lambda_1 B_\gamma g + f, \quad f \in G_1^\perp. \tag{7}$$

It is clear that equations (7) and (1) are equivalent on $G_1^\perp$ because the sum in (5) vanishes if $g \in G_1^\perp$. Therefore every solution $g \in G_1^\perp$ of (7) is a solution of (1) and vice versa.

Theorem 1. *The operator* B_γ *defined in (5) has no characteristic values in the disk* $|\lambda| < r_\gamma$. *If* $|\lambda_1(1 + \gamma\lambda_1)^{-1}| > |\lambda_2|$ *then the iterative process*

$$g_{n+1} = \lambda_1 B_\gamma g_n + F, \quad g_0 = F \equiv \lambda_1 Af - f, \quad F \in G_1^\perp \tag{8}$$

converges no more slowly than a geometric series with ratio q, $0 < q < |\lambda_1\lambda_2^{-1}|$ *to an element* $g = \Phi - f$, *where* $\Phi \in N(I - \lambda_1 A)$ *and* $P\Phi = Pf$. *If* $\dim G_1 = 1$, $\phi \in N(I - \lambda_1 A)$, $\psi \in G_1$ *and* $||\psi|| = ||\phi|| = 1$ *then* $\Phi = \phi(f,\psi)/(\phi,\psi)$. *Process (8) is stable in the following sense: the sequence*

$$h_{n+1} = \lambda_1 B_\gamma h_n + F + \varepsilon_n, \quad h_0 = F, \quad ||\varepsilon_n|| < \varepsilon \tag{9}$$

satisfies the estimate

$$\lim_{n \to \infty} \sup ||g - h_n|| = 0(\varepsilon), \tag{10}$$

where

$$g = \lim_{n \to \infty} g_n. \tag{11}$$

Theorem 2. *If* $\dim G_1 = 1$ *then the iterative process*

$$f_{n+1} = \lambda_1 Af_n, \quad f_0 = f \tag{12}$$

converges no more slowly than a geometrical series with ratio $q = |\lambda_1\lambda_2^{-1}|$ *to the element* $a\phi$, $\phi \in N(I - \lambda_1 A)$, $a = (f,\psi)/(\phi,\psi)$. *Here* $f \in H$ *is arbitrary.*

Proof of Theorem 1: If $g = \lambda B_\gamma g$, then $(g,\psi_j) = \lambda\lambda_1^{-1}(g,\psi_j) + \lambda\gamma(g,\psi_j)$, or $(g,\psi_j)(1 - \lambda\lambda_1^{-1} - \lambda\gamma) = 0$. If for some j, $1 \le j \le m$, $(g,\psi_j) \ne 0$, then $\lambda = \lambda_1(1 + \lambda_1\gamma)^{-1}$. If $(g,\psi_j) = 0$ for all $1 \le j \le m$, then $B_\gamma g = Ag$, $g = \lambda Ag$, i.e., $\lambda \in \sigma_1(A)$, where $\sigma_1(A)$ is the set of all characteristic values of A except λ_1. The value λ_1 is excluded because if $(g,\psi_j) = 0$, $1 \le j \le m$, then $g = 0$, since λ_1 is semisimple. Therefore the disk $|\lambda| < r_\gamma$ does not contain any characteristic values of B_γ. Our argument shows that $\sigma(B_\gamma) \subset \{\sigma\ (A)\} \cup \{\lambda_1(1+\lambda_1\gamma)^{-1}\}$. If $g = \lambda B_\gamma g$, $g \in G_1^\perp$ then $g = \lambda Ag$. Let us show that every $\lambda \in \sigma_1(A)$ belongs to $\sigma(B_\gamma)$. It is sufficient to prove that if $g = \lambda_n Ag$, $n > 1$ then $g \in G_1^\perp$. In order to prove this we start with the identity $(g,\psi_j) = \lambda_n(Ag,\psi_j) = \lambda_n(g,A^*\psi_j) = \lambda_n\lambda_1^{-1}(g,\psi_j)$. Thus $(g,\psi_j)(1-\lambda_n\lambda_1^{-1}) = 0$, $1 \leq j \leq m$. Since $\lambda_n\lambda_1^{-1} \neq 1$ it follows that $g \in G_1^\perp$. We proved that every $\lambda \in \sigma_1(A)$ belongs to $\sigma(B_\gamma)$ and moreover the eigenvectors of A corresponding to λ_n, $n > 1$, are the eigenvectors of B corresponding to λ_n.

Let us prove that process (8) converges. If γ is chosen so that $|\lambda_1(1+\gamma\lambda_1)^{-1}| > |\lambda_2|$ then there are no characteristic values of B_γ in the disk $|\lambda| < |\lambda_2|$. Therefore process (8) converges no more slowly than the geometric series with ratio $0 < q < |\lambda_1\lambda_2^{-1}|$. Since $F \in G_1^\perp$ implies that $AF = B_\gamma F$, one can see that $g \equiv \Sigma_{j=0}^\infty \lambda_1^j B_\gamma^j F = F + \lambda_1 B_\gamma g$ and $B_\gamma g = Ag$. Therefore $g + f = \lambda_1 Af + \lambda_1 B_\gamma g = \lambda_1 Af + \lambda_1 Ag = \lambda_1 A(g+f)$. This means that $h \equiv g + f \in N(I - \lambda_1 A)$. Since $Pg = 0$ we have $Ph = Pf$. If $\dim G_1 = 1$ then $\dim N(I - \lambda_1 A) = 1$. Let $\phi \in N(I - \lambda_1 A)$, $\psi \in G_1$, $||\phi|| = ||\psi|| = 1$. Then $h = c\phi$, $(h,\psi) = c(\phi,\psi)$, i.e., $c = (h,\psi)/(\phi,\psi) = (f,\psi)/(\phi,\psi)$. Note that $(\phi,\psi) \neq 0$ because λ_1 is semisimple, and $(g,\psi) = 0$ because $g \in G_1^\perp$. Let us prove (10). We have

$$h_n = \sum_{j=0}^n (\lambda_1 B_\gamma)^j F + \sum_{j=0}^{n-1} (\lambda_j B_\gamma)^j \epsilon_{n-1-j}, \quad ||\lambda_1 B_\gamma|| \leq q < 1,$$

$$||g-h_n|| \leq \epsilon \sum_{j=0}^{n-1} q^j + \sum_{j=n+1}^\infty q^j ||F|| \leq \frac{\epsilon + ||F||q^{n+1}}{1 - q}.$$

This implies (10). □

Proof of Theorem 2: First let us formulate and prove a lemma.

Lemma 2. Let $f(\lambda)$ be a function of the complex variable λ with values in the set of linear bounded operators on a Banach space. Let $f(\lambda)$ be analytic in the disk $|\lambda| < r$ and meromorphic in the disk $|\lambda| < r + \epsilon$, $\epsilon > 0$. Suppose that λ_1 is a simple pole of $f(\lambda)$, $\mathop{\mathrm{res}}\limits_{\lambda=\lambda_1} f(\lambda) = c$ and $f(\lambda) = \Sigma_{n=0}^\infty a_n\lambda^n$ for $|\lambda| < r$. If there are no other poles in the disk $|\lambda| < r + \epsilon$, then

$$\lim_{n\to\infty} \lambda_1^{n+1} a_n = -c. \tag{13}$$

Proof of Lemma 2: The function $f(\lambda) - c(\lambda-\lambda_1)^{-1}$ is analytic in the disk $|\lambda| < r + \epsilon$. Therefore $f(\lambda) - c(\lambda-\lambda_1)^{-1} = \Sigma_{n=0}^\infty b_n\lambda^n$,

$|\lambda| < r + \varepsilon$. For $|\lambda| < r$ the identity $\Sigma_{n=0}^{\infty} b_n \lambda^n = \Sigma_{n=0}^{\infty} (a_n + c\lambda_1^{-n-1}) \lambda^n$ holds. This identity can be analytically continued into the disk $|\lambda| < r + \varepsilon$. Thus $a_n + c\lambda_1^{-(n+1)} \to 0$ as $n \to \infty$. This implies (13). □

Now we can prove Theorem 2: The function $(I - \lambda A)^{-1} f = \Sigma_{j=0}^{\infty} \lambda^j A^j f$ is analytic in the disk $|\lambda| < |\lambda_1|$, has a simple pole at $\lambda = \lambda_1$, and has no other poles in the disk $|\lambda| < |\lambda_2|$. Lemma 2 says that $\lim_{n \to \infty} \lambda_1^{n+1} A^n f = -c$ with the rate of convergence $O(|\lambda_1 \lambda_2^{-1}|^n)$. Since $f_n = \lambda_1^n A^n f$ we conclude that $\lim_{n \to \infty} f_n = h$ exists and $h = \lambda_1 Ah$. If $\dim N(I - \lambda_1 A) = 1$ then $h = a\phi$, $\phi \in N(I - \lambda_1 A)$. Note that

$$(f_{n+1}, \psi) = \lambda_1 (Af_n, \psi) = (f_n, \psi) = \ldots = (f, \psi).$$

Therefore $a(\phi, \psi) = (f, \psi)$, $a = (f, \psi)/(\phi, \psi)$. □

Remark 2. Process (12) is unstable in the sense that the process

$$h_{n+1} = \lambda_1 A h_n + \varepsilon_n, \quad ||\varepsilon_n|| < \varepsilon, \quad h_0 = f \tag{14}$$

can diverge because $\lambda_1 \in \sigma(A)$, where $\sigma(A)$ is the set of characteristic values of A. Let $\phi \in N(I - \lambda_1 A)$, $\phi = \psi + h$, where $\psi \in G_1$, $h \in G_1^{\perp}$. From $\phi = \lambda_1 A\phi$ it follows that

$$h = \lambda_1 Ah + F, \quad F \equiv \lambda_1 A\psi - \psi, \quad F \in G_1^{\perp}. \tag{15}$$

A stable iterative process for solution of (15) is given in Theorem 1, namely the process (8). In order to use it one must know a basis of G_1. In the case of electrostatics this basis is known explicitly (e.g., $\psi = 1$ in the case of a single conductor). In the general case one can find numerically an approximation to a basis of G_1. If $\{\psi_j\}$, $1 \le j \le m$, is an orthonormal basis of G_1 and $||\psi_{j\varepsilon} - \psi_j|| < \varepsilon$, then the operator $B_{\gamma, \varepsilon} = \lambda_1 A + \gamma \Sigma_{j=1}^m (\cdot, \psi_{j\varepsilon}) \psi_{j\varepsilon}$ has no characteristic values in the disk $|\lambda| < |\lambda_1| + \delta$ where $\delta = \delta(\varepsilon) > 0$ and $\delta(\varepsilon) \to (|\lambda_2| - |\lambda_1|)$ as $\varepsilon \to 0$ provided that γ is chosen so that $|\lambda_1 (1 + \gamma\lambda_1)^{-1}| > |\lambda_2|$. This follows from the uniform convergence $||B_{\gamma, \varepsilon} - B_\gamma|| \to 0$ as $\varepsilon \to 0$.

Remark 3. One can use the following simple but general principle in order to construct a stable iterative process which converges to $\phi \in N(I - \lambda_1 A)$. The principle can be formulated as follows: Suppose that a convergent iterative process for solutions of some equation (*) $Bg = f$ is known for the exact data f. Then it is possible to construct a stable iterative process for solution of this equation with perturbed data f_δ, $||f_\delta - f|| < \delta$. Indeed, let $S_n f$ be the nth approximation of the iterative process. We assume that each S_n is a continuous operator. We have

$$||S_n f_\delta - g|| \le ||S_n f_\delta - S_n f|| + ||S_n f - g||. \tag{16}$$

Here g is the solution of (*). By our assumption

$$||S_n f - g|| \equiv a(n) \to 0 \quad \text{as} \quad n \to \infty \tag{17}$$

and

$$||S_n f_\delta - S_n f|| \equiv b(\delta,n), \quad b(\delta,n) \to 0 \quad \text{as} \quad \delta \to 0. \tag{18}$$

The last limit is not uniform in n. Let us find for any given $\delta > 0$ such $n(\delta)$ that

$$b(\delta,n) + a(n) = \min \equiv \alpha(\delta) \tag{19}$$

It follows from (17), (18) that

$$n(\delta) \to \infty \quad \text{as} \quad \delta \to 0, \quad \alpha(\delta) \to 0 \quad \text{as} \quad \delta \to 0. \tag{20}$$

Therefore

$$||S_{n(\delta)} f_\delta - g|| \to 0 \quad \text{as} \quad \delta \to 0. \tag{21}$$

Let us summarize this observation.

Proposition 1. *If a convergent iterative process for solution of the equation* $Bg = f$ *is known,* $g_n = S_n f$ *is the nth approximation of this process and each operator* S_n *is continuous, then* $||S_{n(\delta)} f_\delta - g|| \to 0$ *as* $\delta \to 0$ *provided that* $n(\delta)$ *is chosen from* (19) *and* $||f_\delta - f|| \le \delta$.

In practice, if B is the linear operator $I - A$, then

$$S_n = \sum_{j=0}^{n} A^j, \quad ||S_n|| \le \frac{||A||^{n+1} - 1}{||A|| - 1}$$

if $||A|| > 1$, and $||S_n|| \le n+1$ if $||A|| \le 1$. This gives an explicit estimate for $b(\delta,n)$ (e.g., if $||A|| \le 1$ then $b(\delta,n) \le \delta(n+1)$). To estimate $a(n)$ one must use specific information about A. For example, under the assumptions of Theorem 2 one has $a(n) \le c|\lambda_1 \lambda_2^{-1}|^n$.

2. The spectral radius of a linear bounded operator A on a Banach space is defined as $r(A) = \lim_{n\to\infty}||A^n||^{1/n}$. This limit always exists [10]. If $|\lambda| > r(A)$ then $(A - \lambda I)^{-1}$ exists and is bounded. Let us assume that

$$r(A) = 1, \quad 1 \notin \sigma(A). \tag{22}$$

It is clear that the equation

$$g = Ag + f \tag{23}$$

is equivalent to the equation

$$g = Bg + f(1+t)^{-1}, \quad t \neq -1, \quad B \equiv (A+tI)(1+t)^{-1}. \tag{24}$$

Consider the iterative process

$$g_{n+1} = Bg_n + f(1+t)^{-1}, \quad g_0 = f(1+t)^{-1}, \quad t > 0. \tag{25}$$

Theorem 3. *The solution* g *of equation* (23) *can be obtained by means of the iterative process* (25), $g = \lim_{n \to \infty} g_n$. *The process converges no more slowly than a convergent geometric series.*

Proof of Theorem 3: The equation

$$g = \lambda Bg + g_0 \tag{26}$$

coincides with (24) if $\lambda = 1$ and can be solved by iterations for sufficiently small $|\lambda|$, $|\lambda| < \delta$. Its solution

$$g(\lambda) = \sum_{n=0}^{\infty} \lambda^n B^n g_0 \tag{27}$$

is analytic in the disk $|\lambda| < \delta$. If $g(\lambda)$ has no singular points in the disk $|\lambda| \leq R$, then the series (27) converges in this disk. If $R > 1$ then the series converges for $\lambda = 1$ at the rate of the geometric series with ratio R^{-1}.

Let us prove that for some $R > 1$ the function $g(\lambda)$ is analytic in the disk $|\lambda| \leq R$. Let us rewrite (26) as

$$g = zAg + bf, \quad z = \frac{\lambda}{1+t-\lambda t}, \quad b = \frac{1}{1+t-\lambda t}. \tag{28}$$

The solution of (28) is analytic in a domain Δ of the complex plane z. This domain includes the disk $|z| < 1$ and a neighborhood of the point $z = 1$. For any $t > 0$ one can find $R > 1$ such that the disk $|\lambda| \leq R$ is mapped by the function $z = \lambda(1+t-\lambda t)^{-1}$ onto a disk $K_r \subset \Delta$. This implies the conclusion of Theorem 3. Indeed, the function $z = \lambda(1+t-\lambda t)^{-1}$ is analytic in the disk $|\lambda| \leq R$, $1 < R < 1+t^{-1}$, and maps this disk onto $K_r \subset \Delta$. The solution $g(z(\lambda))$ to (26) is analytic in the disk $|\lambda| \leq R$. It remains to show that for some $1 < R < 1+t^{-1}$ the function $z = \lambda(1+t-\lambda t)^{-1}$ maps the disk $|\lambda| \leq R$ into Δ. Since $z(\lambda)$ is linear-fractional it maps disks onto disks. Note that $z(\bar{\lambda}) = \overline{z(\lambda)}$ where the bars denote complex conjugation. Therefore the circle $|\lambda| = R$ is mapped onto the circle K_r with the diameter $[z(-R), z(R)]$, $r = [z(R) - z(-R)]/2$ and the center lies on the real axis at the point $[z(R) + z(-R)]/2$. Hence $K_r \subset \Delta$ provided that $z(-R) > -1$, $|z(R) - 1| < \alpha$ where $\alpha > 0$ is sufficiently small. We have $z(-R) = -R(1+t+Rt)^{-1} > -1$ if $t > (R-1)/(R+1)$; or, if $t > 0$ is fixed, $z(-R) > -1$ when $R < (1+t)/(1-t)$ for $t < 1, z(-R) > -1$ for any $R > 0$ when $t \geq 1$. On the other hand $z(R) = R(1+t-tR)^{-1} < 1 + \alpha$ if $R < (1+\alpha)(1+t)/(1+t(1+\alpha)) = 1 + \alpha[1 + t(1+\alpha)]^{-1}$. Therefore there exists $R > 1$ which satisfies the last inequality. We proved that for some $1 < R < 1+t^{-1}$ the function $z(\lambda)$ maps the disk $|\lambda| \leq R$ onto the disk $K_r \subset \Delta$. This completes the proof of Theorem 3. One can choose $t > 0$ so that R will be maximal and the rate of convergence of the process (25) will be maximal in this case. □

3. Let us formulate a well-known theorem [12]. Let A be a linear bounded operator on a Banach space X and $\sigma(A)$ be its characteristic set (i.e., the image of the spectrum of A under the mapping $z \to z^{-1}$).

Theorem 4. If $\sigma(A) \subset \{\lambda: |\lambda| > 1\}$ *then for every* $f \in X$ *the equation*

$$g = Ag + f \tag{29}$$

has a unique solution g*, given by the iterative process*

$$g_{n+1} = Ag_n + f, \quad g = \lim_{n \to \infty} g_n \tag{30}$$

for any initial approximation g_0*. If there are points of* $\sigma(A)$ *in the disk* $|\lambda| < 1$ *then there exists a set* $E \subset X$ *such that* E *is of the second category and the process* (30) *diverges if* $f \in E$ *and* $g_0 = 0$*.*

The set E is said to be of the second category if it is not a countable union of nowhere dense sets.

§2. Iterative Processes for Solving Some Operator Equations

Let A be a selfadjoint linear operator on a Hilbert space H, $||A|| = 1$. Consider the equation

$$g = Ag + f. \tag{1}$$

The following theorem is proved in [16].

Theorem 1. Suppose that -1 *is not a characteristic value of* A *and* (1) *is solvable. Then the iterative process*

$$g_{n+1} = Ag_n + f, \tag{2}$$

converges to a solution of (1) *for any* $g_0 \in H$.

Proof of Theorem 1 ([16]): Let H_1 be the eigenspace of A corresponding to $\lambda = 1$ and let P_1 be the projection on H_1. If g is a solution to (1) then $g' = g - P_1 g$ is also a solution to (1) and $g' \perp H_1$. Let us prove that $g_n \to g' + P_1 g_0$ as $n \to \infty$. Let $0 < \delta < 1$ and $P_2 = \int_{-1+\delta}^{1-\delta} dE_\lambda$, $P_3 = I - P_1 - P_2$, where $A = \int_{-1}^{1} dE_\lambda$ is the spectral representation of A. The operator P_3 is an orthoprojection and since -1 is not a characteristic value of A, one has

$$||P_3 f|| \to 0 \quad \text{as} \quad \delta \to 0 \quad \text{for any fixed} \quad f \in H. \tag{3}$$

Since $AP_j = P_j A$ and $P_i P_j = 0$ for $i \neq j$ one can rewrite (2) as

$$P_1 g_{n+1} = AP_1 g_n + P_1 f, \tag{4}$$

$$P_2 g_{n+1} = AP_2 g_n + P_2 f, \tag{5}$$

$$P_3 g_{n+1} = AP_3 g_n + P_3 f. \tag{6}$$

Since (1) is solvable, $P_1 f = 0$ and (4) shows that $P_1 g_n = P_1 g_0$. Let $H_2 = P_2 H$. The process (5) can be considered as an iterative process for the restriction A_2 of A to H_2. Since $||A_\varepsilon|| \leq 1-\delta$ the process (5) converges to $P_2 g'$ which is the solution to the equation $h = Ah + P_2 f$. Thus $||P_2 g_n - P_2 g'|| < \varepsilon$ for $n > n(\varepsilon)$. Furthermore,

$$||P_3(g_n - g')|| = ||A(P_3 g_{n-1} - P_3 g')|| \leq ||P_3(g_{n-1} - g')||$$
$$\leq \ldots \leq ||P_3(g_0 - g')|| < \varepsilon$$

provided that δ is sufficiently small (see (3)).

Now one has

$$||g_n - (g' + P_1 g_0)|| \leq ||P_1(g_n - g' - g_0)|| + ||P_2(g_n - g')||$$
$$+ ||P_3(g_n - g')|| < 2\varepsilon$$

provided that $n > n(\varepsilon)$ and δ is sufficiently small. This completes the proof.

The following result is also discussed in [16].

Theorem 2. *Every solvable linear equation with a bounded operator in a Hilbert space can be solved by an iterative process.*

Proof of Theorem 2: Let

$$Bg = f \tag{7}$$

and let B be a linear bounded operator. The equation

$$Ag \equiv B*Bg = B*f \tag{8}$$

is equivalent to (7). Indeed, (7) implies (8). On the other hand, since (7) is solvable $f = Bh$ and (8) can be written as $B*B(g-h) = 0$. Multiplying this by $g - h$ yields $B(g-h) = 0$, i.e., $Bg = Bh = f$. That is, (8) implies (7). Equation (8) can be written as

$$g = (I - kA)g + F, \quad F = kB*f, \tag{9}$$

where $k > 0$ is a constant. Suppose that

$$0 < k < 2||A||^{-1}. \tag{10}$$

Then the operator $I - kA$ is selfadjoint, -1 is not an eigenvalue of it, and $||I - kA|| \leq 1$. By Theorem 1, equation (9) is solvable by the iterative process

$$g_{n+1} = (I - kA)g_n + F, \tag{11}$$

with an arbitrary initial element $g_0 \in H$. □

Remark 1. If $0 < m \leq A \leq M$ then

$$||I - kA|| = \min = \frac{M-m}{M+m} \quad \text{if} \quad k = \frac{2}{m+M} .$$

Indeed $||I - kA|| = \max_{m \leq \lambda \leq M}|1-k\lambda|$. Since $\max_{m \leq \lambda \leq M}|1-k\lambda| = \max(|1-km|, |1-kM|)$ its minimum is attained when $|1-km| = |1-kM|$, i.e., $k = 2(m+M)^{-1}$. This minimum is $(M-m)/(M+m)$.

This is also a well-known observation.

Remark 2. Let $B \geq 0$ be a linear operator on a Hilbert space H such that equation (7) is solvable. Then the iterative process

$$g_{n+1} + Bg_{n+1} = g_n + f, \tag{12}$$

converges to a solution of (7) for any initial element $g_0 \in H$.

Proof of Remark 2 [28u, p. 78]: We have $g_{n+1} = (I+B)^{-1}g_n + (I+B)^{-1}f$. For the operator $A = (I+B)^{-1}$ the assumptions of Theorem 1 hold and Remark 2 follows from this theorem. □

§3. Iterative Processes for Solving the Exterior and Interior
      Boundary Value Problems

1. Consider the problems

$$\Delta u = 0 \quad \text{in} \quad D, \quad u|_\Gamma = f, \tag{1}$$

$$\Delta v = 0 \quad \text{in} \quad D_e, \quad \frac{\partial v}{\partial N_e}\Big|_\Gamma = f, \quad v(\infty) = 0, \tag{2}$$

$$\Delta u = 0 \quad \text{in} \quad D_e, \quad u|_\Gamma = f, \quad u(\infty) = 0, \tag{3}$$

$$\Delta v = 0 \quad \text{in} \quad D, \quad \frac{\partial v}{\partial N_i} = f, \quad \int_\Gamma f\,dt = 0. \tag{4}$$

Define

$$v = \int_\Gamma \frac{\sigma(t)dt}{4\pi r_{xt}}, \quad u = \int_\Gamma \mu(t)\,\frac{\partial}{\partial N_t}\,\frac{1}{4\pi r_{xt}}\,dt. \tag{5}$$

It is known that

$$u_{\substack{i\\e}} = \frac{A^*\mu \bar{+} \mu}{2}, \quad A^*\mu = \int_\Gamma \mu(t)\,\frac{\partial}{\partial N_t}\,\frac{1}{2\pi r_{st}}\,dt, \tag{6}$$

$$\frac{\partial v}{\partial N_{\substack{i\\e}}} = \frac{A\sigma \pm \sigma}{2}, \quad A\sigma = \int_\Gamma \sigma(t)\,\frac{\partial}{\partial N_s}\,\frac{1}{2\pi r_{st}}\,dt, \tag{7}$$

and $\partial u/\partial N_i = \partial u/\partial N_e$ provided that Γ is smooth [8]. In (6) and (7) the upper (lower) signs correspond to the upper (lower) subscript $i(e)$.
    Therefore from (1), (2), and (5)-(7) it follows that

$$\mu = A^*\mu - 2f, \tag{8}$$

$$\sigma = A\sigma - 2f. \tag{9}$$

It is well known 8 that A and A^* have no characteristic values in the disk $|\lambda| < 1$ and only the characteristic value $\lambda = -1$ on the circle $|\lambda| = 1$. The operators A and A^* are compact in $C(\Gamma)$ and $H = L^2(\Gamma)$ if Γ is smooth. We assume that D is a bounded domain in R^3, D_e is the exterior domain, and their boundary Γ is a smooth closed surface. Therefore we have

Proposition 1. *Theorem 6.1.3 is applicable to equations (8) and (9).*

Remark 1. Setting $t = 1$, $B = A^*$ in (1.25) yields the classical Neumann process for solving the interior Dirichlet problem which reduces to equation (8).

    From (4)-(7) it follows that the problem (4) can be reduced to the integral equations

$$\sigma = -A\sigma + 2f, \qquad \int_\Gamma f dt = 0. \tag{10}$$

Equation (10) was discussed in detail in Chapter 2. Theorem 6.1.1 was basic in this discussion and the crucial assumption (1.2) is fulfilled for the operator A defined in (7) and $\lambda_1 = -1$. Actually in this case λ_1 is simple, i.e., dim $N(I+A) = 1$ and the function $\psi = 1$ belongs to $N(I+A^*) = G_1$. Condition (10) means that $f \in G_1^\perp$. Therefore equation (1) can be solved by the iterative process

$$\sigma_{n+1} = -A\sigma_n + 2f, \qquad \sigma_0 = 2f \tag{11}$$

or its modification (1.8) which guarantees the stability of the calculations with respect to small errors.

2. Let us discuss problem (3). The usual equation $\mu = -A^*\mu + 2f$ can have no solutions and is not equivalent to problem (3) because the solution to (3) is not necessarily of the form (5). Therefore we look for a solution to (3) of the form

$$u = \frac{a}{|x|} + \int_\Gamma \mu(t) \frac{\partial}{\partial N_t} \frac{1}{4\pi r_{xt}} dt, \qquad a = \text{const.} \tag{12}$$

From (12) and (13) it follows that

$$\mu = -A^*\mu + 2(f - \frac{a}{|s|}), \qquad s = x|_\Gamma. \tag{13}$$

Consider the equation

$$\nu = M\nu + 2(f - \frac{a}{|s|}), \qquad M\nu \equiv -A^*\nu + \int_\Gamma \nu dt. \tag{14}$$

Proposition 2. *The operator M has no characteristic values in the disk $|\lambda| \leq 1$, so that the iterative process*

$$\nu_{n+1} = M\nu_n + 2(f - \frac{a}{|s|}), \tag{15}$$

converges (in $C(\Gamma)$) to the solution of equation (14) for an arbitrary $\nu_0 \in C(\Gamma)$. Moreover, one can choose a so that equation (14) and (13) are equivalent, i.e., so that

$$\int_\Gamma \nu dt = 0. \tag{16}$$

This will be true if

$$a = \int_\Gamma Qf \, dt \left(\int_\Gamma Q \left(\frac{1}{|s|}\right) ds \right)^{-1}, \qquad Q \equiv (I - M)^{-1}. \tag{17}$$

Proof of Proposition 2: First let us prove that the disk $|\lambda| \leq 1$ contains no characteristic values of M. Let

$$\nu = \lambda M\nu = -\lambda A^*\nu + \lambda \int_\Gamma \nu dt, \tag{18}$$

and

$$u(x) = \int_\Gamma \nu(t) \frac{\partial}{\partial N_t} \frac{1}{4\pi r_{xt}} dt. \tag{19}$$

Then from (18), (19), and (6) it follows that

$$(1+\lambda)u_e = (1-\lambda)u_i + \lambda \int_\Gamma (u_e - u_i)dt. \tag{20}$$

Multiplying (20) by $(\overline{\partial u/\partial N}) = (\overline{\partial u/\partial N_e}) = (\overline{\partial u/\partial N_i})$ one obtains

$$\frac{1+\lambda}{1-\lambda} \int_\Gamma u_e \frac{\overline{\partial u}}{\partial N_e} dt = \int_\Gamma u_i \frac{\overline{\partial u}}{\partial N_i} dt + \frac{\lambda}{1-\lambda} \int_\Gamma (u_e - u_i)dt \int_\Gamma \frac{\overline{\partial u}}{\partial N} dt. \tag{21}$$

By Green's formula

$$\int_\Gamma u_e \frac{\overline{\partial u}}{\partial N_e} dt = - \int_{D_e} |\nabla u|^2 dx \leq 0, \qquad \int_\Gamma u_i \frac{\overline{\partial u}}{\partial N_i} dt = \int_D |\nabla u|^2 dx \geq 0,$$

$$\int_\Gamma \frac{\partial u}{\partial n} dt = 0. \tag{22}$$

From (22) and (21) it follows that $(1+\lambda)(1-\lambda)^{-1} \leq 0$. Hence λ is real and $|\lambda| \geq 1$. It remains to prove that $\lambda = \pm 1$ is not a characteristic value of M. If $\lambda = -1$ then (21) shows that

$$\int_\Gamma u_i \frac{\overline{\partial u}}{\partial N_i} dt = \int_D |\nabla u|^2 dx = 0. \tag{23}$$

Therefore u is constant in D, $\partial u/\partial N_i = 0 = \partial u/\partial N_e$. Hence $u = 0$ in D_e and $\nu = u_e - u_i = $ const. Without loss of generality, suppose $\nu = 1$ is a solution to (18):

$$1 = A^*1 - S, \qquad S = \text{meas } \Gamma. \tag{24}$$

Let ν_0 be the electrostatic density, i.e.,

$$\nu_0 = -A\nu_0, \qquad \int_\Gamma \nu_0 dt > 0. \tag{25}$$

Multiplying (24) by ν_0 and integrating over Γ one obtains

$$\int_\Gamma \nu_0 dt = (\nu_0, A^*1) - S \int_\Gamma \nu_0 dt,$$

or

$$(1+S) \int_\Gamma v_0 dt = - \int_\Gamma v_0 dt. \tag{26}$$

This is a contradiction. Therefore $v = 1$ is not a solution to (18). If $\lambda = 1$ then $v = -A^*v + \int_\Gamma v dt$. The solvability condition is

$$\int_\Gamma v dt \int_\Gamma v_0 dt = 0. \tag{27}$$

Thus

$$\int_\Gamma v dt = 0 \tag{28}$$

and

$$v = -A^*v, \tag{29}$$

so that $v = \text{const} \neq 0$. This contradicts (28). Therefore $\lambda = 1$ is not a characteristic value of M. The other statements of Proposition 2 are obvious. $\square$

Remark 2. In practice in order to find a from formula (17) one can use the processes

$$h_{n+1} = Mh_n + \frac{1}{|s|}, \quad \lim_{n \to \infty} h_n = Q\left(\frac{1}{|s|}\right) \tag{30}$$

and

$$v_{n+1} = Mv_n + f, \quad \lim_{n \to \infty} v_n = Q(f) \tag{31}$$

and then find a from (17).

3. Consider the third boundary value problem

$$\Delta u = 0 \quad \text{in} \quad D_e, \quad -\frac{\partial u}{\partial N_e} + hu\big|_\Gamma = f, \quad u(\infty) = 0, \tag{32}$$

$$\Delta u = 0 \quad \text{in} \quad D, \quad \frac{\partial u}{\partial N_i} + hu\big|_\Gamma = f, \tag{33}$$

$$h = h_1 + ih_2, \quad h_1 \geq 0, \quad h_2 \leq 0, \quad |h_1| + |h_2| > 0. \tag{34}$$

It is easy to prove that under the assumption (34) problems (32) and (33) have at most one solution.

Let us look for the solution of (32) and (33) of the form

$$u = \int_\Gamma \frac{g(t)dt}{4\pi r_{xt}}. \tag{35}$$

Then problem (32) is reduced to the equation

$$g = Ag - Tg + 2f, \tag{36}$$

where A is defined in (7) and

$$Tg = h \int_{\Gamma} \frac{g(t)dt}{2\pi r_{st}} \equiv hT_1 g, \tag{37}$$

The problem (33) is reduced to the equation

$$g = -Ag - Tg + 2f. \tag{38}$$

Consider the problem

$$g + Tg = \lambda Ag. \tag{39}$$

Theorem 1. *If (34) holds then all the eigenvalues of (39) satisfy the inequality* $|\lambda| > 1$ *and they are real if* $h > 0$. *Moreover the equation*

$$g + Tg = \lambda Ag + F, \qquad \lambda = \pm 1 \tag{40}$$

can be solved by the iterative process

$$g_{n+1} + Tg_{n+1} = \lambda Ag_n + F, \tag{41}$$

where $g_0 \in H = L^2(\Gamma)$ *is arbitrary. This method converges no more slowly than a convergent geometric series.*

Remark 3. The iterative process

$$g_{n+1} + Tg_n = Ag_n + F, \tag{42}$$

with an arbitrary $g_0 \in H$ converges if $0 < h < k$, where

$$k \equiv \min\left\{\int_D |\nabla u|^2 dx \left(\int_{\Gamma} |u|^2 dt\right)^{-1}\right\}. \tag{43}$$

Proof of Theorem 1: Let us rewrite (39) as

$$(1-\lambda) \frac{\partial u}{\partial N_i} + 2hu = (1+\lambda) \frac{\partial u}{\partial N_e}, \tag{44}$$

where u is defined in (35). Multiplying (44) by $\bar{u}$ and integrating over Γ yields

$$\frac{1-\lambda}{1+\lambda} A + h \frac{B}{1+\lambda} = C, \tag{45}$$

where

$$A = \int_{\Gamma} \frac{\partial u}{\partial N_i} \bar{u} dt > 0, \qquad B = 2 \int_{\Gamma} |u|^2 dt > 0, \tag{46}$$

$$C = \int_{\Gamma} \frac{\partial u}{\partial N_e} \bar{u} dt < 0. \tag{47}$$

If A, B, or C is zero then $u \equiv 0$. Let $\lambda = a + ib$. Taking real and imaginary parts of (45) yields

$$\frac{(1-a^2-b^2)A + [h_1(1+a) + h_2 b]B}{(1+a)^2 + b^2} = C < 0,$$ (48)

and

$$\frac{-2bA + [h_2(1+a) - h_1 b]B}{(1+a)^2 + b^2} = 0.$$ (49)

Hence

$$(1-|\lambda|^2)A + [h_1(1+a) + h_2 b]B < 0,$$ (50)

$$h_2 = \frac{h_1 B + 2A}{1 + a} b.$$ (51)

Suppose that $|\lambda| \leq 1$. Since $h_2 \leq 0$, $h_1 \geq 0$, and $|a| \leq |\lambda| \leq 1$, it follows from (51) that $b \leq 0$. Thus $h_2 b > 0$. Therefore (50) cannot be valid. This contradiction proves that $|\lambda| > 1$. If $h = h_1 > 0$, $h_2 = 0$, then $b = 0$, i.e., all the eigenvalues are real-valued. In order to prove that the process (41) converges let us consider the equation (*) $g = \lambda Gg + q$, where $G \equiv (I+T)^{-1}A$, $q = (I+T)^{-1}F$. The operator $(I+T)^{-1}$ exists and is bounded because T is compact and $(I+T)f = 0$ implies that $f = 0$. The latter conclusion follows immediately from the positive definiteness of the operator $Re(I+T) = I + h_1 T_1$, $T_1 > 0$, $h_1 \geq 0$. The operator G has no characteristic values in the disk $|\lambda| \leq 1$ (as was proved above). Therefore the iterative process

$$g_{n+1} = Gg_n + q,$$ (52)

with an arbitrary $g_0 \in H$, converges at the rate of a geometric series to the solution of the equation (*). The process (52) is equivalent to (41) and equation (40) is equivalent to (*). Theorem 1 is proved. □

Proof of Remark 3: Consider the equation

$$g = \mu(-T+A)g.$$ (53)

If the characteristic values $|\mu_j| > 1$ then the process (42) converges. Let us find when $|\mu_j| > 1$. Let us rewrite (53) as

$$(1-\mu) \frac{\partial u}{\partial N_i} + 2\mu h u = (1+\mu) \frac{\partial u}{\partial N_e}.$$ (54)

From (54) it follows that

$$(1-\mu)A + \mu h B = (1+\mu)C,$$ (55)

where A, B, C are defined in (45).

If h > 0 then as in the proof of Theorem 1 one can show that if
μ = a + ib then

$$(1-a)A + ahB = (1+a)C, \tag{56}$$

$$-bA + bhB = bC. \tag{57}$$

If b ≠ 0 then from (57) and (58) it follows that A = C. This is a contradiction because of (46), (47). Thus b = 0, μ = a, and

$$\frac{1-a}{1+a} A + \frac{a}{1+a} hB < 0. \tag{58}$$

Suppose that |a| < 1. Then (58) cannot hold for 0 ≤ a ≤ 1. If
-1 < a < 0 then (58) can be written as

$$\frac{\int_D |\nabla u|^2 dx}{\int_\Gamma |u|^2 dt} < \frac{2|a|h}{1-|a|} \frac{1-|a|}{1+|a|} = \frac{2|a|h}{1+|a|} < h. \tag{59}$$

Since |a| < 1, one has 2|a|h/(1+|a|) < h. Therefore (59) cannot hold
if k > h. If a = -1 then (56) shows that

$$2A = hB, \quad \int_D |\nabla u|^2 dx \left(\int_\Gamma |u|^2 dt \right)^{-1} = h. \tag{60}$$

If k > h the equality (6) cannot hold. This argument proves that if
k > h > 0 then the process (42) converges.

If h > k then equation (56) does not lead to a contradiction even
if |a| < 1. Therefore it is conceivable that the process (42) diverges
for some F. □

4. Consider the problem

$$\Delta u = 0 \quad \text{in} \quad D, \quad u\big|_{\Gamma_1} = f_1, \quad \frac{\partial u}{\partial N_i}\bigg|_{\Gamma_2} = f_2, \quad \Gamma_1 \cup \Gamma_2 = \Gamma, \tag{61}$$

$\Gamma_1 \cap \Gamma_2 = \emptyset, \Gamma_1 \neq \emptyset$, where ∅ denotes the empty set. This problem was
studied probably for the first time by Zaremba (1910). It has at most
one solution. Numerical approaches to this problem have been studied re-
cently by many authors and by means of various techniques (see [36] and
the bibliography in this paper). In this section a simple approach taken
from [28s] is discussed. Consider the problem

$$\Delta v_h = 0 \quad \text{in} \quad D, \quad \frac{\partial v_h}{\partial N_i} + h(s)v_h\big|_\Gamma = F, \tag{62}$$

$$F = \begin{cases} hf_1 & \text{on } \Gamma_1 \\ f_2 & \text{on } \Gamma_2 \end{cases}, \quad h(s) = \begin{cases} h & \text{on } \Gamma_1 \\ 0 & \text{on } \Gamma_2 \end{cases}, \quad h = \text{const} > 0. \tag{63}$$

The idea is to first solve (62) by an iterative process and then to show that $v_h \to u$ as $h \to +\infty$ and establish the estimates

$$||u - v_h||_{H_1} \leq ch^{-1}, \quad ||u - v_h||_{\tilde{H}_2} \leq ch^{-1}. \tag{64}$$

Here and below $c > 0$ denotes various constants, $H_1 = W_2^{\frac{1}{2}}$ is the Sobolev space [12], and $\tilde{H}_2 = W_2^2(\tilde{D})$, where $\tilde{D} \subset D$ is any fixed strictly inner subdomain of D, i.e., $\text{dist}(\text{cl } \tilde{D}, \partial D) > 0$ where ∂D is the boundary of D and $\text{cl } \tilde{D}$ is the closure of $\tilde{D}$.

Theorem 2. *The solution of (62) exists, is unique, and satisfies (64), where* u *is the solution of (61). Furthermore the solution of (62) can be calculated by means of the iterative process described in Theorem 1.*

Proof of Theorem 2: Let $w_h = v_h - u$. Then

$$\Delta w_h = 0 \text{ in } D, \quad \frac{\partial w_h}{\partial N}\Big|_{\Gamma_2} = 0, \quad \frac{\partial w_h}{\partial N} + hw_h\Big|_{\Gamma_1} = -\frac{\partial u}{\partial N}\Big|_{\Gamma_1}.$$

From this it follows that

$$\int_\Gamma w_h \frac{\partial w_h}{\partial N} dt + h \int_{\Gamma_1} |w_h|^2 dt = -\int_{\Gamma_1} w_h \frac{\partial u}{\partial N} dt.$$

Therefore

$$\int_D |\nabla w_h|^2 dx + h \int_{\Gamma_1} |w_h|^2 dt \leq c||w_h||_{L^2(\Gamma_1)}, \quad c = ||\frac{\partial u}{\partial N}||_{L^2(\Gamma_1)}.$$

Thus

$$||w_h||_{L^2(\Gamma_1)} \leq ch^{-1}, \quad \int_D |\nabla w_h|^2 dx \leq c^2 h^{-1}. \tag{65}$$

From (65) and the inequality

$$||w_h||_{L^2(D)} \leq C_1(||\nabla w_h||_{L^2(D)} + ||w_h||_{L^2(\Gamma_1)}) \tag{66}$$

where $C_1 = C_1(D, \Gamma_1)$, the first estimate (63) follows. The second estimate (64) follows from the inequality

$$||w||_{\tilde{H}_2} \leq C_2(||\Delta w||_{L^2(D)} + ||w||_{L^2(D)})$$

which is valid for any function $w \in W_2^2(D)$ and any $\tilde{D} \subset D$ which is a strictly inner subdomain of D [12].

It remains to prove that the problem (62) can be solved by an iterative process.

To this end one can use a generalization of Theorem 1. Define T as in (37) with h = h(s) where h(s) is defined in (63) or 0 < m ≤ h(s) ≤ M is a piecewise continuous function. The conclusion and the proof of Theorem 1 remain valid. The only new point in the proof is the invertibility of the operator I + T. This new point is separated as the following lemma. □

Lemma 1. Under the above assumptions on h(s), the operator $(I+T)^{-1}$ is bounded and defined on all of $H = L^2(\Gamma)$.

Proof of Lemma 1: Since T is compact it is sufficient to prove that (*) f + Tf = 0 implies f = 0. If 0 < m ≤ h(s) ≤ M and $h^{-\frac{1}{2}}f = g$ then g + Sg = 0, where $S = h^{\frac{1}{2}}T_1 h^{\frac{1}{2}}$ and T_1 is defined in (37). Therefore S ≥ 0 and I + S ≥ I. Thus g = 0 and f = 0. If h(s) is defined in (63) then (*) shows that f = 0 on Γ_2 and

$$f(s) + h \int_{\Gamma_1} \frac{f(t)dt}{2\pi r_{st}} = 0, \quad s \in \Gamma_1, \quad h > 0 \tag{67}$$

Since the kernel r_{st}^{-1} is positive semidefinite, (67) implies that f = 0 on Γ_1. This completes the proof. □

§4. **An Iterative Process for Solving the Fredholm Integral Equations of the First Kind with Pointwise Positive Kernel**

In Sec. 2.4 a problem of practical interest was discussed, reduced to equation (2.4.1), and solved by means of the iterative process (2.4.2). Here we give a theoretical justification of this process in a general setting.

Consider the equation

$$Kf = \int_D k(x,y)f(y)dy = g(x), \quad x \in D \subset R^r, \tag{1}$$

where D is a bounded domain, the operator $K: L^2(D) \to L^2(D)$ is compact and

$$K(x,y) > 0 \tag{2}$$

almost everywhere. Suppose there exists a function h(x) > 0 such that Kh ≤ c and $\int_D a(x)dx < \infty$, where $a(x) \equiv h(x)/(Kh(x))$. Let $\phi = fa^{-1}(x)$ and $H_{\pm} = L^2(D, a^{\pm 1}(x))$, $||f||_{\pm}^2 = \int_D |f|^2 a^{\pm 1}(x)dx$. Let us rewrite (1) as

$$K_1\phi = g, \quad K_1\phi \equiv \int_D K(x,y)a(y)\phi(y)dy = K_1 a\phi. \tag{3}$$

Let

$$Q = I - K_1, \quad K_1 f_j = \lambda_j f_j, \quad \lambda_1 > |\lambda_2| \geq |\lambda_3| \geq \cdots \qquad (4)$$

The first eigenvalue of the integral operators with pointwise positive kernels is positive and simple, i.e., the corresponding eigenspace is one-dimensional (Perron-Frobenius theorem for matrices, Jentzsch theorem for integral operators, Krein-Rutman theorem for abstract operators [37].
Let us assume that

$$g(x) \in H_+, \qquad (5)$$

$$0 < c_1(\Delta) \leq \int_\Delta K(x,y)a(y)dy \leq c_2(\Delta), \quad x \in D, \qquad (6)$$

where $\Delta \subset D$, meas $\Delta > 0$,

equation (3) is solvable in H_+, $\qquad (7)$

the eigenfunctions $\{f_j\}$ form a Riesz basis of H_+, $\qquad (8)$

$$|\arg \lambda_j| \leq \frac{\pi}{3}, \quad \lambda_j \neq 0. \qquad (9)$$

Theorem 1. *If the above assumptions (2)-(9) hold then the iterative process*

$$\phi_{n+1} = Q\phi_n + g, \quad \phi_0 = g \qquad (10)$$

converges in H_+ *to a solution* ϕ *of (3). The function* $f = a\phi$ *is a solution to (1),* $f \in H_-$.

Remark 1. A complete minimal system $\{f_j\} \subset H$ forms a Riesz basis of the Hilbert space H if for any numbers $c_1,\ldots,c_n$ and any n the inequality

$$a \sum_{j=1}^{n} |c_j|^2 \leq ||\sum_{j=1}^{n} c_j f_j||^2 \leq b \sum_{j=1}^{n} |c_j|^2, \quad a > 0 \qquad (11)$$

holds, where a, b do not depend on n.

Proof of Theorem 1: Let ϕ be a solution to (3), $g_n = \phi - \phi_n$.
Then $g_n = Q^n g$. Let $g = \sum_{j=1}^{\infty} c_j f_j$. Then

$$g_n = \sum_{j=1}^{\infty} (1-\lambda_j)^n c_j f_j, \text{ and } |\lambda_j| < 1 \text{ if } j \geq 2.$$

From (9) it follows that $|1-\lambda_j| < 1$. Indeed, if $\lambda = r \exp(i\psi)$, $r < 1$, $|\psi| \leq \pi/3$, then $|1-\lambda|^2 = 1 + r^2 - 2r \cos \psi \leq 1 + r^2 - r < 1$. Hence

$|1-\lambda_j|^n \to 0$ as $n \to \infty$. Therefore $||g_n||^2 \le b \sum_{j=1}^{\infty} |1-\lambda_j|^{2n} |c_j|^2 \to 0$ as $n \to \infty$. This means that $||\psi_n - \phi||_{H_+} \to 0$ as $n \to \infty$. The rest is obvious. □

Example. Let $\Gamma = \{x: |x| = 1\}$, $m = 2$. Equation (1) is of the form

$$Af = \int_{-\pi}^{\pi} \ln \left| \frac{1}{2 \sin \frac{\phi - \phi'}{2}} \right| f(\phi') d\phi' = g(\phi), \quad -\pi \le \phi \le \pi. \tag{12}$$

Since $\int_0^{\pi} \ln \sin x \, dx = -\pi \ln 2$ one has

$$\int_{-\pi}^{\pi} \ln \left| \frac{1}{2 \sin \frac{\phi - \phi'}{2}} \right| d\phi' = 0.$$

Therefore $f_0 = (2\pi)^{-1}$ is the solution of the homogeneous equation (12). In this example equation (12), if solvable, is equivalent to the equation

$$Bf = -\int_{-\pi}^{\pi} \ln |\sin\{(\phi-\phi')/2\}| f(\phi') d\phi' = g(\phi), \quad -\pi \le \phi \le \pi \tag{13}$$

with the pointwise positive and selfadjoint kernel, provided that one looks for a solution of (12) which satisfies the condition $\int_0^{2\pi} f \, du = 0$. In this example $a(x) = (2\pi \ln 2)^{-1} \equiv a$, $B_1 = aB$, $f = a\psi$ and (10) takes the form

$$\begin{cases} \psi_{n+1}(\phi) = \psi_n(\phi) + (2\pi \ln 2)^{-1} \int_{-\pi}^{\pi} \ln |\sin\{(\phi-\phi')/2\}| \psi_n(\phi') d\phi' + g(\phi) \\ \psi_0 = g(\phi). \end{cases} \tag{14}$$

Let $g(\phi) = \cos \phi$. Since

$$-\ln \left| \sin \frac{\phi'-\phi}{2} \right| = \ln 2 + \sum_{m=1}^{\infty} \frac{\cos\{m(\phi'-\phi)\}}{m} \tag{15}$$

one has

$$B \cos \phi = \pi \cos \phi; \quad -B_1 \cos \phi = -(2 \ln 2)^{-1} \cos \phi. \tag{16}$$

With this in mind one concludes from (14) that

$$\psi_1 = \cos \phi \, (2-(\ln 4)^{-1}) = c_1 \cos \phi,$$

$$\psi_2 = (1+c_1) \cos \phi - c_1 (\ln 4)^{-1} \cos \phi \equiv c_2 \cos \phi,$$

and

$$\psi_{n+1} = c_{n+1} \cos \phi, \quad c_{n+1} = (1 + c_n) - c_n (\ln 4)^{-1}. \tag{17}$$

Thus

$$c_{n+1} = qc_n + 1, \quad c_0 = 1, \quad q = 1 - (\ln 4)^{-1} = 0.28. \tag{18}$$

Therefore

$$\lim c_n = c = \ln 4 = 2 \ln 2,$$

$$\psi = \lim \psi_n = 2 \ln 2 \cos \phi, \quad |\psi - \psi_n| \leq (1-q)^{-1} q^{n+1},$$

$$f = \pi^{-1} \cos \phi.$$

Chapter 7. Wave Scattering by Small Bodies

§1. Introduction

Wave scattering by small bodies is of great interest in theory and
for applications. An incomplete list of problems for which wave scattering by small bodies is of prime importance includes: radio wave scattering by rain and hail, light scattering by cosmic dust, light scattering
in colloidal solutions, light propagation in muddy water, wave scattering
by small inhomogeneities in a medium. We will show that the skin effect
for thin wires and radiation from small holes are also dependent on wave
scattering by small bodies. The number of examples is practically un-
limited. The theory was originated by Rayleigh (1871) who contributed to
this field until his death (1919). Rayleigh understood that the main term
in the scattering amplitude in the problem of wave scattering by a small
body with diameter much less than the wavelength of the exterior field
can be described as dipole radiation. J. J. Thomson (1893) realized that
for a small perfect conductor the magnetic dipole radiation is of the same
order as the electric dipole radiation. Some efforts were made in order
to develop an algorithm for finding the expansion of the scattered field
in powers of ka, where k is the wave number and a is the characteristic dimension of the scatterer, ka ≪ 1 (Stevenson [31], Kleinman [14]).
Since in many cases the first term of this expansion already provides a
good approximation we will only discuss this first approximation. The
general idea of our presentation is very simple. First it will be shown
that a low-frequency approximation to the scattering matrix can be calculated if the electric and magnetic polarizability tensors for the scatterer
are known. This result is known, but the explicit formula for the scat-

tering matrix is new. Explicit approximate analytical formulas for the polarizability tensors were obtained in Chapter 4. Therefore we have explicit approximate analytical formulas for the scattering matrix. Using these formulas one can produce computer programs for calculating the scattering matrix for a small body of an arbitrary shape. Senior and Ahlgren (see Bibliographical notes) prepared a computer program for calculating the elements of the polarizability tensor of rotationally symmetric perfect conductors. Exact solutions in closed form for the exterior problems of potential theory in three dimensions case are known for ellipsoids only.

The other important point which sould be emphasized is that we study dependence of the scattering matrix on the boundary condition.

§2. Scalar Wave Scattering: The Single-Body Problem

1. Consider the problem

$$(\nabla^2 + k^2)v = 0 \quad \text{in} \quad \Omega, \tag{1}$$

$$\frac{\partial v}{\partial N} - hv\Big|_\Gamma = \left(-\frac{\partial u_0}{\partial N} + hu_0\right)\Big|_\Gamma, \tag{2}$$

$$|x|\left(\frac{\partial v}{\partial |x|} - ikv\right) \to 0 \quad \text{as} \quad |x| \to \infty, \tag{3}$$

where u_0 is the incident field, Ω is the exterior domain with smooth boundary Γ, $h = $ const, $h = h_1 + ih_2$, $h_2 \le 0$, $h_1 \ge 0$, $k > 0$, $D = R^3 \backslash \Omega$ is the interior bounded domain. Let us look for a solution of (1)-(3) of the form

$$v(x) = \int_\Gamma g(x,s,k)\sigma(s)ds, \quad g = \frac{\exp(ik|x-s|)}{4\pi|x-s|}. \tag{4}$$

The scattering amplitude $f(n,k)$ is defined from the formula

$$v \sim \frac{\exp(ik|x|)}{|x|} f(n,k), \quad |x| \to \infty, \quad n = x|x|^{-1}. \tag{5}$$

From (4) and (5) it follows that

$$f(n,k) = (4\pi)^{-1} \int_\Gamma \exp\{-ik(n,s)\}\sigma(s)ds. \tag{6}$$

Substituting (4) into (2) yields

$$\sigma = A(k)\sigma - hT(k)\sigma - 2hu_0 + 2\frac{\partial u_0}{\partial N}, \tag{7}$$

where

$$A(k)\sigma = 2\int_{\Gamma} \frac{\partial}{\partial N_s} g(s,t,k)\sigma(t)dt, \tag{8}$$

$$T(k)\sigma = 2\int_{\Gamma} g(s,t,k)\sigma(t)dt. \tag{9}$$

Let us expand σ, $A(k)$, $T(k)$, and u_0 in powers of k.

$$\sigma = \sigma_0 + ik\sigma_1 + \frac{(ik)^2}{2}\sigma_2 + \cdots , \tag{10}$$

$$A(k) = A + ikA_1 + \frac{(ik)^2}{2}A_2 + \cdots , \tag{11}$$

$$T(k) = T + ikT_1 + \frac{(ik)^2}{2}T_2 + \cdots , \tag{12}$$

$$u_0 = u_{00} + iku_{01} + \frac{(ik)^2}{2}u_{02} + \cdots . \tag{13}$$

From (10)-(13) and (7) it follows that

$$\sigma_0 = A\sigma_0 - hT\sigma_0 - 2hu_{00} + 2\frac{\partial u_{00}}{\partial N} , \tag{14}$$

$$\sigma_1 = A\sigma_1 - hT\sigma_1 + A_1\sigma_0 - hT_1\sigma_0 - 2hu_{01} + 2\frac{\partial u_{01}}{\partial N} , \tag{15}$$

$$\sigma_2 = A\sigma_2 - hT\sigma_2 + A_2\sigma_0 + 2A_1\sigma_1 - hT_2\sigma_0 - 2hT_1\sigma_1$$
$$- 2hu_{02} + 2\frac{\partial u_{02}}{\partial N} . \tag{16}$$

From (15) and (6) it follows that

$$4\pi f = \int_{\Gamma}\left[1 - ik(n,s) + \frac{(ik)^2}{2}(n,s)^2 + \cdots\right]\left[\sigma_0 + ik\sigma_1 + \frac{(ik)^2}{2}\sigma_2 + \cdots\right]ds$$

$$= \int_{\Gamma}\sigma_0 ds + ik\left[\int_{\Gamma}\sigma_1 ds - \int_{\Gamma}(n,s)\sigma_0 ds\right]$$

$$+ \frac{(ik)^2}{2}\left[\int_{\Gamma}\sigma_2 ds - 2\int_{\Gamma}(n,s)\sigma_1 ds + \int_{\Gamma}\sigma_0(n,s)^2 ds\right] + \cdots \tag{17}$$

Let us assume that

$$u_0 = \exp\{ik(\nu,x)\}. \tag{18}$$

Then

$$u_{00} = 1, \quad u_{01} = (\nu,s), \quad u_{02} = (\nu,s)^2 \tag{19}$$

$$\frac{\partial u_{00}}{\partial\partial N} = 0, \quad \frac{\partial u_{01}}{\partial N} = (\nu,N), \quad \frac{\partial u_{02}}{\partial N} = 2(\nu,s)(\nu,N).$$

We note that the following formulas hold

$$A\sigma = \int_\Gamma \frac{\partial}{\partial N_s} \frac{1}{2\pi r_{st}} \sigma(t)dt, \quad A_1\sigma = 0, \tag{20}$$

$$-\int_\Gamma A\sigma dt = \int_\Gamma \sigma dt, \tag{21}$$

$$T\sigma = \int_\Gamma \frac{\sigma dt}{2\pi r_{st}}, \tag{22}$$

$$T_1\sigma = (2\pi)^{-1} \int_\Gamma \sigma(t)dt. \tag{23}$$

Let us integrate (14) over Γ and take into account (19)-(21). This yields

$$2\int_\Gamma \sigma_0 dt = -2hS - h \int_\Gamma\int_\Gamma \frac{\sigma_0(t)dtds}{2\pi r_{st}}, \quad S = \text{meas } \Gamma,$$

or

$$\int_\Gamma \sigma_0 dt = -hS - \frac{h}{4\pi} \int_\Gamma\int_\Gamma \frac{\sigma_0(t)dtds}{r_{st}}. \tag{24}$$

The exact value of σ_0 should be found from the integral equation (14). An approximate value of $\int_\Gamma \sigma_0 dt$ can be found from (24) if one uses the approximation

$$\int_\Gamma \frac{ds}{r_{st}} \approx \frac{1}{S} \int_\Gamma dt \int_\Gamma \frac{ds}{r_{st}} = JS^{-1}, \quad J \equiv \int_\Gamma\int_\Gamma r_{st}^{-1}dsdt. \tag{25}$$

From (25) and (24) it follows that

$$\int_\Gamma \sigma_0 dt \approx -\frac{hS}{1+hJ(4\pi S)^{-1}}. \tag{26}$$

In Chapter 3 the approximate formula

$$C \approx C^{(0)} = 4\pi S^2 J^{-1}, \quad \varepsilon_0 = 1 \tag{27}$$

was given. Combining (26) and (27) yields

$$\int_\Gamma \sigma_0 dt \approx -hS(1 + hSC^{-1})^{-1}. \tag{28}$$

Therefore

$$f(n,k) \approx -hS(1 + hSC^{-1})^{-1} \frac{u_{00}}{4\pi}. \tag{29}$$

If $h = \infty$, i.e., the scatterer is a perfect conductor, then

$$f = \frac{-C}{4\pi} u_{00}. \tag{30}$$

This result is well known, but its derivation differs from the usual [8].

From (29) and (30) it follows that the scattering from a small body of arbitrary shape under the Dirichlet boundary condition (i.e., acoustically soft body, $h = \infty$) or under the impedance boundary condition ($h < \infty$) is isotropic and the scattering amplitude is of order a, where a is the characteristic dimension of the scatterer. Note that if the scatterer is not too prolate then $C \sim a$. We also assumed above that h is not too small, e.g., $hS > c^{-1}$.

3. Consider now the case when $h = 0$, i.e., the case of the acoustically rigid body. We shall see that in this case the scattering is anisotropic, is defined by the magnetic polarizability tensor and the scattering amplitude is of order $k^2 a^3$. If $h = 0$ then (14) takes the form $\sigma_0 = A\sigma_0$ and therefore $\sigma_0 = 0$ since 1 is not an eigenvalue of A. Equation (15) takes the form

$$\sigma_1 = A\sigma_1 + 2 \frac{\partial u_{01}}{\partial N}. \tag{31}$$

Integrating (31) over Γ and using (21) yields

$$\int_{\Gamma} \sigma_1 dt = \int_{\Gamma} \frac{\partial u_{01}}{\partial N} dt = \int_{D} \Delta u_{01} dx = 0, \tag{32}$$

since $u_{00} = 0$ and $u_{01} = 0$. The latter equations follow from the equation

$$(\Delta + k^2)u_0 = 0$$

and the asymptotic expansion (13).

Thus, in the case $h = 0$ formula (17) takes the form

$$4\pi f(n,k) = -\frac{k^2}{2} \int_{\Gamma} \sigma_2 ds + k^2 \int_{\Gamma} (n,s)\sigma_1 ds. \tag{33}$$

For the initial field (18) it follows from (31) that

$$\left(n, \int_{\Gamma} s\sigma_1 ds\right) = -V\beta_{pq}\nu_q n_p. \tag{34}$$

Here and below one should sum over the repeating indices, V denotes the volume of the scatterer and β_{pq} is the magnetic polarizability tensor defined in Chapter 5 as $V\beta_{pq} = V^{-1} \int_{\Gamma} s_p \sigma_q(s) ds$, where σ_q is the solution of the equation $\sigma_q = A\sigma_q - 2N_q$ and N is the unit outer normal to Γ. In order to calculate the term $\int_{\Gamma} \sigma_2 ds$ let us rewrite (16) for $h = 0$ as

$$\sigma_2 = A\sigma_2 + 2\frac{\partial u_{02}}{\partial N}. \tag{35}$$

Here we used (20) and took into account that $\sigma_0 = 0$. From (19) and (35) it follows that

$$\sigma_2 = A\sigma_2 + 4(\nu,s)(\nu,N). \tag{36}$$

Integrating (36) over Γ and taking into account (21) yields

$$\int_\Gamma \sigma_2 dt = 2\int_\Gamma (\nu,s)(\nu,N)ds = 2\left(\nu, \int_\Gamma N(\nu,s)ds\right)$$

$$= 2\left(\nu, \int_D \nabla(\nu,x)dx\right) = 2(\nu,\nu)V = 2V. \tag{37}$$

From (33), (34), and (37) it follows that if $h = 0$ and the initial field is given by (18) then the scattering amplitude

$$f(n,\nu,k) = -\frac{k^2V}{4\pi} - \frac{k^2V}{4\pi}\beta_{pq}\nu_q n_p, \qquad f \sim k^2 a^3. \tag{38}$$

The scattering is anisotropic in this case.

4. Let us derive the following formula for the scattering amplitude in the case $h = 0$ for an arbitrary initial field u_0:

$$f(n,k) = \frac{ikV}{4\pi}\beta_{pq}\frac{\partial u_0}{\partial x_q}n_p + \frac{V\Delta u_0}{4\pi}. \tag{39}$$

The main assumption is the smallness of the scatterer. We want to derive formula (39) for two reasons. First, the initial field u_0 is not assumed to be a plane wave. Second, we want to isolate the dependence of the scattering amplitude on the size of the body from its dependence on the wave number k. When the initial field is $u_0 = \exp\{ik(\nu,x)\}$ and the scatterer is placed at the origin, then the small parameter is ka, so that $k \to 0$ is equivalent to $x \to 0$. But if we consider the many-body problem then the phase difference should be taken into account. For example, if k is small but x is large then $ik(\nu,x)$ is not necessarily small, and such a situation occurs in the many-body problem if the distance between some of the bodies is larger than the wavelength.

As in Sec. 3 we consider the problem (1)-(3) with $h = 0$ and look for the solution of the form (4). The integral equation for σ takes the form (7) with $h = 0$. If a is very small we can rewrite this equation as

$$\sigma = A\sigma + 2\frac{\partial u_0}{\partial N}, \tag{40}$$

where A is defined in (20) and the error is $O(a)$.

Let us rewrite formula (6) as

$$f(n,k) = (4\pi)^{-1} \int_{\Gamma} \sigma(s)ds - ik(4\pi)^{-1} \int_{\Gamma} (n,s)\sigma(s)ds. \tag{41}$$

Here we took into account that $ka \ll 1$. We expand the initial field u_0 in a Taylor series assuming that the origin is placed inside the scatterer. This yields

$$u_0(x,k) = u_{00} + (v,x) + \frac{1}{2}(Bx,x) + O(a^3) \tag{42}$$

where $u_{00} = u_0(0,k)$,

$$v = \Delta u_0(x,k)\big|_{x=0}, \quad (B)_{mj} = b_{mj} = \frac{\partial^2 u_0(x,k)}{\partial x_m \partial x_j}\bigg|_{x=0} \tag{43}$$

Therefore

$$\frac{\partial u_0}{\partial N} = (v,N) + (Bs,N), \tag{44}$$

where $s \in \Gamma$. Integrating (40) over Γ and taking into account (21), one obtains

$$\int_{\Gamma} \sigma ds = \int_{\Gamma} (Bs,N)ds = V \operatorname{tr} B = V\Delta u_0\big|_{x=0}, \tag{45}$$

where tr is the trace. Furthermore, one obtains

$$-ik(4\pi)^{-1} n_p \int_{\Gamma} \sigma s_q(s)ds = ik(4\pi)^{-1} n_p v_q \beta_{pq} V, \tag{46}$$

where $\beta_{pq} = \beta_{qp}$ is the magnetic polarizability tensor defined in Chapter 5 as

$$V\beta_{pq} = \int_{\Gamma} s_q \sigma_p(s)ds, \tag{47}$$

where σ_p is the solution of the equation

$$\sigma_p = A\sigma_p - 2N_p. \tag{48}$$

Formula (39) follows from (45), (46), and (43). In calculating the integral in (46) one can neglect the term (Bs,N) in the right-hand side of (44) because this term is of order $O(a)$, while $(v,N) = O(1)$.

§3. Scalar Wave Scattering: The Many-Body Problem

1. Consider scattering by r bodies. Let

$$D = \bigcup_{j=1}^{r} D_j, \quad \Gamma = \bigcup_{j=1}^{r} \Gamma_j, \quad D_j \cap D_i = \emptyset, \quad i \neq j, \quad \Omega = R^3 \diagdown D, \qquad (1)$$

where $\emptyset$ denotes the empty set, $R^3 \diagdown D$ denotes the complement of D in R^3, and Γ_j is the boundary of D_j. Let

$$h\big|_{\Gamma_j} = h_j = h_{1j} + ih_{2j}, \quad h_{1j} \geq 0, \quad h_{2j} \leq 0, \quad |h_j| > 0, \qquad (2)$$

$$a = \max_{1 \leq j \leq r} a_j, \qquad (3)$$

$$d = \min_{i,j} d_{ij}, \quad i \neq j \qquad (4)$$

$$\ell = \max_{i,j} d_{ij}, \quad \int_{\Gamma} \equiv \sum_{j=1}^{r} \int_{\Gamma_j}, \qquad (5)$$

where d_{ij} is the distance between D_i and D_j.

Consider the problem (2.1)-(2.3), which looks formally identical in the cases $r = 1$ and $r > 1$. As in §2 we look for a solution of the form (2.4) and define the scattering amplitude by formula (2.5). The scattering amplitude can be written as in (2.6), $\sigma = (\sigma_1, \ldots, \sigma_r)$ but an important difference between cases $r = 1$ and $r > 1$ is that if $r = 1$ then $|s| \sim a$ in (2.6), while if $r > 1$ the magnitude $|s|$ can be large.

Let us denote by s_j some point inside D_j and rewrite (2.6) for the case $r > 1$ as

$$f(n,k) = (4\pi)^{-1} \sum_{j=1}^{r} \int_{\Gamma_j} \exp\{-ik(n,s-s_j)\} \sigma_j(s) ds \, \exp\{-ik(n,s_j)\}. \qquad (6)$$

In (6) the magnitudes $|s-s_j| \sim a$ if $s \in \Gamma_j$ and $|s-s_j| \sim d_{ij}$ if $s \in \Gamma_i$. The integral equations for σ_j, $1 \leq j \leq r$, can be obtained by substituting (2.4) into the boundary condition (2.2). This yields

$$\sigma_j = A_j(k)\sigma_j - h_j T_j(k)\sigma_j + \sum{}' A_{jp}(k)\sigma_p - h_j \sum{}' T_{jp}(k)\sigma_p$$

$$+ 2\frac{\partial u_0}{\partial N} - 2h_j u_0, \quad 1 \leq j \leq r, \quad \sum{}' = \sum_{p=1, p \neq j}^{r}, \qquad (7)$$

where

$$A_{jp}\sigma_p = \int_{\Gamma_p} \frac{\partial}{\partial N_{s_j}} \frac{\exp(ikr_{s_jt_p})}{2\pi r_{s_jt_p}} \sigma_p(t_p)dt_p, \tag{8}$$

$$T_{jp}\sigma_p = \int_{\Gamma_p} \frac{\exp(ikr_{s_jt_p})}{2\pi r_{s_jt_p}} \sigma_p(t_p)dt_p. \tag{9}$$

Suppose that

$$d \gg a. \tag{10}$$

If one neglects the terms A_{jp} and T_{jp} for $j \neq p$ in (7) then for σ_j one obtains the same equations as for a single body in §2. Therefore for $h_j \neq 0$ the scattering amplitude can be calculated from the formula

$$f(n,k) = -\frac{1}{4\pi} \sum_{j=1}^{r} \exp\{-ik(n,s_j)\} \frac{h_j S_j}{1+h_j S_j C_j^{-1}} u_{0j}, \tag{11}$$

where C_j is the capacitance of the jth body, S_j is the area of its sur-face, $u_{0j} = u_0(s_j,k)$ (see formula (2.29)). If we assume that every small body is affected by the self consistent field u in the medium consisting of many small bodies, then (11) takes the form

$$f(n,k) = -\frac{1}{4\pi} \int \exp\{-ik(n,y)\}q(y)u(y,k)dy, \tag{12}$$

where $q(y)$ is the "effective potential" which is defined as

$$q(y) = N(y) \frac{hS}{1 + hSC^{-1}}. \tag{13}$$

Here $N(y)$ is the number of the small bodies (particles) per unit volume and $hS(1 + hSC^{-1})$ is the average value of $h_j S_j (1 + h_j S_j C_j^{-1})^{-1}$ in a neighborhood of the point y. The integral in (12) is taken over the do-main where $N(y) \neq 0$. The self-consistent field u satisfies the equa-tion

$$u(x,k) = u_0(x,k) - \int \frac{\exp(ik|x-s|)}{4\pi|x-s|} q(y)u(y,k)dy, \tag{14}$$

which is obtained by taking the limit $r \to \infty$ in the formula

$$u(x,k) = u_0(x,k) - \sum_{j=1}^{r} \frac{\exp(ik|x-s_j|)}{4\pi|x-s_j|} \frac{h_j S_j}{1 + hSC^{-1}} u(s_j,k). \tag{15}$$

These arguments are of a heuristic nature only. Equation (14) can be written as the Schrödinger equation

$$[\nabla^2 + k^2 - q(x)]u(x,k) = 0, \tag{16}$$

$$u - u_0 \sim \frac{\exp(ik|x|)}{4\pi|x|} \, f(n,k) \quad \text{as} \quad |x| \to \infty. \tag{17}$$

2. If the number r of the small scatterers is not very large
$(r \sim 10)$ then the scattering amplitude and the scattered field can be
found from a linear system of algebraic equations. The matrix of the system has dominant main diagonal so that the system can be easily solvable
by iterations. In order to prove this statement let us look for the solution of the problem (2.1)-(2.3) (with $\Gamma = \cup_{j=1}^{r} \Gamma_j$) of the form

$$v = \sum_{j=1}^{r} \int_{\Gamma_j} \frac{\exp(ik|x-s|)}{4\pi|x-s|} \, \sigma_j(s)ds. \tag{18}$$

In general, in order to find v one derives a system of integral equations for finding σ_j, $1 \le j \le r$. In our case when $ka \ll 1$ the scattering amplitude can be calculated as

$$
\begin{aligned}
f(n,k) &= \frac{1}{4\pi} \sum_{j=1}^{r} \int_{\Gamma_j} \exp\{-ik(n,s)\}\sigma_j(s)ds \\
&= \frac{1}{4\pi} \sum_{j=1}^{r} \exp\{-ik(n,s_j)\}Q_j + O(ka),
\end{aligned} \tag{19}
$$

where

$$Q_j = \int_{\Gamma_j} \sigma_j(t)dt, \tag{20}$$

and we assume that $Q_j \ne 0$, $1 \le j \le r$. This is the case when $h_j \ne 0$.
Consider, for example, the Dirichlet boundary condition ($h_j = \infty$) (acoustically soft particles). Then from (18) and (2.2) it follows that

$$\int_{\Gamma_m} \frac{\exp(ik|x_m-s|)}{4\pi|x_m-s|} \, \sigma_m(s)ds + \sum_{m \ne j, j=1}^{r} \int_{\Gamma_j} \frac{\exp(ik|x_m-s|)}{4\pi|x_m-s|} \, \sigma_j ds$$

$$= -u_0(x_m,k), \quad 1 \le m \le r. \tag{21}$$

If $ka \ll 1$, then this system can be written with the accuracy $O(ka)$ as

$$\int_{\Gamma_m} \frac{\sigma_m ds}{4\pi|x_m-s|} = -u_0(x_m,k) - \sum_{m \ne j, j=1}^{r} \frac{\exp(ik|x_m-s_j|)}{4\pi|x_m-s_j|} \, Q_j,$$

$$1 \le m \le r. \tag{22}$$

Equation (22) can be considered as an equation for the electrostatic charge
distribution σ_m on the surface Γ_m of the perfect conductor charged to

the potential given by the right-hand side of (22). Therefore the total charge on Γ_m is

$$Q_m = \int_{\Gamma_m} \sigma_m ds = C_m \left\{ -u_0(x_m,k) - \sum_{m \neq j, j=1}^{r} \frac{\exp(ik|x_m-s_j|)}{4\pi|x_m-s_j|} Q_j \right\},$$

where C_m is the electrical capacitance of the perfect conductor with boundary Γ_m. The above system of equations for $Q = (Q_1,\ldots,Q_r)$ can be written as

$$AQ = b, \tag{23}$$

where

$$A = (a_{mj}) = \delta_{mj} + C_m \frac{\exp(ik|x_m-s_j|)}{4\pi|x_m-s_j|}, \qquad b_m = -C_m u_0(x_m,k),$$

$$\delta_{mj} = \begin{cases} 1, & m = j, \\ 0, & m \neq j. \end{cases} \tag{24}$$

If the particles are not too prolate then $C_m \sim a$. The matrix A will have dominant main diagonal if

$$(4\pi)^{-1} rda < 1, \tag{25}$$

where d is defined in (4). If condition (25) holds then the system (23) can be solved by iterations and the scattering amplitude can be found from formula (19). The scattered field v can be found from the formula

$$v = \sum_{j=1}^{r} \frac{\exp(ik|x-s_j|)}{4\pi|x-s_j|} Q_j \tag{26}$$

with the accuracy $O(ka)$. If $h \neq 0$ then the scattering amplitude can be calculated from (19) and (20), and the linear algebraic system for Q_m can be obtained from (7). To this end let us integrate (7) over Γ_j, yielding

$$Q_j = -Q_j - \frac{h_j J_j}{2\pi S_j} Q_j + \sum' d_{jp} Q_p - 2h_j S_j u_0(s_j). \tag{27}$$

Here we used arguments similar to those given in §2, subsection 2, and the following notations:

$$J_j = \int_{\Gamma_j}\int_{\Gamma_j} \frac{dsdt}{r_{st}}, \qquad d_{jp} = \int_{\Gamma_j} ds \left\{ \frac{\partial}{\partial N_s} \frac{\exp(ikr_{st_p})}{2\pi r_{st_p}} - h_j \frac{\exp(ikr_{st_p})}{2\pi r_{st_p}} \right\}.$$

Equation (27) can be written as

$$\tilde{A}Q = \tilde{b}, \tag{28}$$

$$(\tilde{A}_{jp}) = \tilde{a}_{jp} = \delta_{jp}\left(1 + \frac{h_j J_j}{4\pi S_j}\right)-\tilde{d}_{jp}, \quad \tilde{d}_{jp} = \frac{d_{jp}}{2}, \quad \tilde{b}_j = -h_j S_j u_0(s_j). \quad (29)$$

The linear system (28) can be solved by iterations if

$$1 + \frac{h_j J_j}{4\pi S_j} > \sum_{p \neq j, p=1}^{r} |\tilde{d}_{jp}|, \quad 1 \leq j \leq r. \quad (30)$$

If $h_j = 0$, $1 \leq j \leq r$, then $Q_j = 0$ and formula (19) for the scattering amplitude becomes more complicated. This was shown in §2. If we consider each of the small bodies as being affected by a self-consistent field u, then from (19) and (2.39) it follows that

$$f(n,k) = (4\pi)^{-1} \sum_{j=1}^{r} \exp\{-ik(n,s_j)\}\left\{ikV_j \beta_{pq}^{(j)} \frac{\partial u}{\partial x_q} n_p + V\Delta u\right\}, \quad (31)$$

where V_j is the volume of the jth body and $\beta_{pq}^{(j)}$ is its magnetic polarizability tensor. The same argument leads to the following formula for the self-consistent field in the medium:

$$u = u_0 + \sum_{j=1}^{r} \frac{\exp(ik|x-s_j|)}{4\pi|x-s_j|}\left\{ikV_j \beta_{pq}^{(j)} \frac{\partial u(s_j,k)}{\partial x_q} n_p + V_j \Delta u(s_j,k)\right\}, \quad (32)$$

where s_j is the radius vector of the jth body.

If one passes to the limit as $r \to \infty$ in (32) then the integro-differential equation for the field u takes the form

$$u(x,k) = u_0(x,k) + \int \frac{\exp(ik|x-y|)}{4\pi|x-y|}\left(ikB_{pq}(y) \frac{\partial u}{\partial x_q} \frac{x_p - y_p}{|x-y|}\right.$$
$$\left. + b(y)\Delta u\right)dy, \quad (33)$$

where one must sum over the repeating indices, the integral is taken over the domain where $b(y) \neq 0$, $b(y)$ is the average volume of the bodies near the point y, and $B_{pq}(y)$ is the average magnetic polarizability tensor. That is, if $K_h(y)$ is the ball of radius h centered at y, then

$$b(y) = \lim_{h \to 0} \frac{\Sigma V_j}{|K_h(y)|}, \quad B_{pq}(y) = \lim_{h \to 0} \frac{\Sigma V_j \beta_{pq}^{(j)}}{|K_h(y)|}, \quad (34)$$

where $|K_h(y)|$ is the volume of $K_h(y)$ and Σ denotes the sum over the bodies which are in the ball $K_h(y)$. The vector $(x_p-y_p)/|x-y|$ in formula (33) replaces n_p in formula (32).

§4. Electromagnetic Wave Scattering

1. Let us consider the scattering by a single homogeneous body D with characteristic dimension a. Let ε, μ, σ be its dielectric permeability, magnetic permeability, and conductivity, ε_0, μ_0, $\sigma_0 = 0$ be the corresponding parameters of the exterior medium, $\varepsilon' = \varepsilon + i\sigma\omega^{-1}$, ω be the frequency of the initial field, λ_0 be its wavelength and $k_0 = 2\pi\lambda_0^{-1}$. Let $\lambda = \lambda_0(|\varepsilon'\mu|)^{-\frac{1}{2}}$ be the wavelength in the body, and $\delta = \{2(|\varepsilon'\mu|\omega^2)\}^{\frac{1}{2}}$ be the depth of the skin layer. We consider scattering under the following assumptions, which will be treated separately:

$$|\varepsilon'| \gg 1, \quad \delta \gg a, \quad k_0 a \ll 1, \tag{1}$$

$$|\varepsilon'| \gg 1, \quad \delta \ll a, \quad k_0 a \ll 1, \tag{2}$$

$$|(\varepsilon'-\varepsilon_0)\varepsilon_0^{-1}| + |(\mu-\mu_0)\mu_0^{-1}| \ll 1. \tag{3}$$

Assumption (1) corresponds to a small dielectric body. Assumption (2) corresponds to a small well-conducting body. Assumption (3) corresponds to the case when the body does not differ much from the exterior medium and does not require the body to be small. Our aim is to derive explicit analytical approximate formulas for the scattering amplitude and for the scattering matrix.

2. The basic equations are

$$\text{curl } E = i\omega\mu H, \quad \text{curl } H = -i\omega\varepsilon'E \quad \text{in } D, \tag{4}$$

$$\text{curl } E = i\omega\mu_0 H, \quad \text{curl } H = -i\omega\varepsilon_0 E + j_0 \quad \text{in } \Omega, \tag{5}$$

where Ω is the exterior domain with respect to D. The boundary conditions are

$$N \times E \quad \text{and} \quad \mu H \cdot N \quad \text{are continuous when crossing } \Gamma, \tag{6}$$

where Γ is the boundary of D and N is the outward pointing unit normal at the boundary.

If $\sigma = \infty$ then

$$N \times E = 0 \quad \text{on } \Gamma. \tag{6'}$$

This case can occur only under assumption (2). In (5), j_0 is the initial current source. Let

$$A_0 = \int G(x,y)j_0 dy, \quad G = \frac{\exp(ik|x-y|)}{4\pi|x-y|}, \quad k_0^2 = \omega^2\varepsilon_0\mu_0, \quad \int = \int_{R^3} \tag{7}$$

and

$$E_0 = \frac{1}{-i\omega\epsilon_0} (\text{curl curl } A_0 - j_0), \quad H_0 = \text{curl } A_0, \tag{8}$$

The total field can be found from the formulas

$$E = E_0 + E_1, \quad H = H_0 + H_1, \tag{9}$$

where

$$E_1 = \frac{1}{-i\omega\epsilon_0} \text{curl curl } A - \text{curl } F, \tag{10}$$

$$H_1 = \frac{1}{-i\omega\epsilon_0} \text{curl curl } F + \text{curl } A, \tag{11}$$

and

$$A = \int_\Gamma G(x,s)N \times H_1 ds, \quad F = -\int_\Gamma G(x,s)N \times E_1 ds. \tag{11}$$

Remark 1. If one tries under the assumption (1) to calculate the scattering using the approximations $n \times E_1 = -N \times E_0$ on Γ and $N \times H_1 = 0$ on Γ then one can see that this leads to wrong results (for example one can take the spherical scatterer D and use the explicit solution which is known for this case). Therefore the above approximations, which are used in geometrical optics, are not good in our low-frequency problem.

3. Let $n = x|x|^{-1}$, $|x| = r$ and

$$f = f_E(n,k) = \lim_{|x| \to \infty, x|x|^{-1} = n} x \exp(-ik|x|)E_1. \tag{12}$$

Let us prove under the assumptions (2) that

$$f = \frac{k_0^2}{4\pi\epsilon_0} [n,[P,n]] + \frac{k_0^2}{4\pi} \left(\frac{\mu_0}{\epsilon_0}\right)^{\frac{1}{2}} [M,n], \tag{13}$$

where P and M are the electric and magnetic dipole moments induced on the body by the initial field and $[A,B] = A \times B$ is the vector product. Let us consider the first two terms of the expansion of the vector potential in powers of ka.

$$\begin{aligned}
A &= (4\pi)^{-1} \int_D j(y)G(x,y)dy \\
&= \frac{\exp(ik_0|x|)}{4\pi|x|} \int_D dy j(y) \exp\{-ik_0(n,y)\} \\
&= \frac{\exp(ik_0|x|)}{4\pi|x|} \left\{\int_D j(y)dy - ik_0 \int_D (n,y)j(y)dy + \cdots\right\} \\
&= \frac{\exp(ik_0|x|)}{4\pi|x|} \{-i\omega P - ikM \times n\}, \tag{14}
\end{aligned}$$

where

$$P = \int_D y\mu(y)dy, \quad M = \frac{1}{2}\int_D [y,j]dy, \tag{15}$$

and $\rho = (i\omega)^{-1}\text{div } j$. Indeed

$$-i\omega P = -i\omega \int_D y\rho(y)dy = -\int_D y \text{ div } jdy = -\int_\Gamma (j,N)yds + \int_D jdy$$

$$= \int_D jdy, \tag{16}$$

because $(j,N) = 0$ on Γ. Furthermore,

$$\int_D (n,y)jdy = \frac{1}{2}\int_D ([y,j] \times n + j(n,y) + y(n,j))dy = M \times n, \tag{17}$$

because

$$\int_D (j(n,y) + y(n,j))dy + \int_D y(n,y)\text{div } jdy = \int_\Gamma y(j,N)(y,n)ds = 0, \tag{18}$$

and the integral

$$\int_D y(n,y)\text{div } jdy = i\omega \int_D y(n,y)\rho dy \approx 0.$$

In the far-field zone, $j = 0$ and $E_1 = (-i\omega\varepsilon_0)^{-1}$ curl curl A. Therefore from (14) and (12) it follows that

$$f = -(4\pi i\omega\varepsilon_0)^{-1} ik_0 n \times [ik_0 n \times \{-i\omega P - ik_0 M \times n\}]$$

$$= \frac{k_0^2}{4\pi\varepsilon_0} n \times [P,n] + \frac{k_0^2}{4\pi}\left(\frac{\mu_0}{\varepsilon_0}\right)^{\frac{1}{2}} M \times n. \tag{19}$$

If the domain D shrinks to a surface S then (19) still holds with

$$P = \int_S s\sigma(s)ds, \quad M = \frac{1}{2}\int_S s \times j \ ds. \tag{20}$$

Algorithms and formulas for calculating P and M are given in Chapter 5.

Under the assumption (1) the magnetic dipole radiation can be neglected if $\mu = \mu_0$ because the eddy currents are negligible if $\delta \gg a$. Under the assumption (2) the magnetic dipole radiation is of the order of the electric dipole radiation even in the case $\mu = \mu_0$.

In general, the magnetic polarizability vector can be calculated from the formula

$$M_i = \tilde{\beta}_{ij}V\mu_0 H_j, \tag{21}$$

where

$$\tilde{\beta}_{ij} = \alpha_{ij}(-1) + \alpha_{ij}(\gamma_\mu), \qquad \gamma_\mu = \frac{\mu - \mu_0}{\mu + \mu_0} \tag{22}$$

and $\alpha_{ij}(\gamma)$ is defined in §5.1. We denote

$$\alpha_{ij}(-1) = \beta_{ij}. \tag{23}$$

If $\mu = \mu_0$ then $\alpha_{ij}(\gamma_\mu) = 0$.

Remark 2. Suppose that D is a metallic body. In this case the current $j = N \times H$, where H is the magnetic field on the surface of the body. Let H^1 denote the magnetic field on the surface Γ of the ideal magnetic insulator D, i.e., a body with $\mu = 0$. This field is the value on Γ of the solution of the problem

$$\text{curl } H = 0, \quad \text{div } H = 0 \quad \text{in } \Omega, \quad N \cdot H = 0 \quad \text{on } \Gamma, \quad H(\infty) = H^0, \tag{24}$$

where H^0 is a given constant field. In the quasistatic problem H^0 is the initial field at the point where the small body is placed. If $\delta \ll a$ then neither magnetic nor electric field can penetrate into the body and therefore the body behaves like a perfect magnetic insulator in the initial homogeneous magnetic field H^0. Under the assumption (2) a good approximation for $N \times H$ is $N \times H^1$. This approximation leads to the correct value of M. On the other hand, this approximation leads to a wrong value of P. Let us show it in the case when D is a ball of radius a. The magnetic field H^1 in this case is known:

$$H^1 = H^0 - \frac{a^3}{2|x|^3} \left\{ 3 \frac{x(x, H^0)}{|x|^2} - H^0 \right\}. \tag{25}$$

Therefore

$$N \times H^1 = \frac{3}{2} \frac{[x, H^0]}{|x|}, \qquad A \times B = [A, B], \tag{26}$$

and

$$-i\omega P = \int_\Gamma j \, dy = \int_\Gamma N \times H^1 dy = \frac{3}{2a} \int_\Gamma [x, H^0] ds = 0,$$

which is wrong. Thus one can calculate M using the approximation

$$j = N \times H^1 \tag{27}$$

if the body is metallic, but this approximation cannot be used for calculating P.

4. Let us calculate f under the assumption (3). Equations (4) and (5) can be written as

$$\text{curl } E = i\omega\mu_0 H + i\omega(\mu-\mu_0)\eta H, \tag{28}$$

$$\text{curl } H = -i\omega\varepsilon_0 E + j_0 - i\omega(\varepsilon'-\varepsilon_0)\eta E \tag{29}$$

where

$$\eta = \begin{cases} 1, & x \in D, \\ 0, & x \notin D. \end{cases} \tag{30}$$

Let us set

$$j_e = -i\omega(\varepsilon'-\varepsilon_0)\eta E, \quad j_m = -i\omega(\mu-\mu_0)\eta H, \tag{31}$$

$$A = \int G(x,y)j_e dy, \quad F = \int G(x,y)j_m dy, \quad \int = \int_D. \tag{32}$$

Then the vectors E_1, H_1 defined in formula (9) can be found from the formulas

$$E_1 = -(i\omega\varepsilon_0)^{-1}(\text{curl curl } A - j_e) - \text{curl } F, \tag{33}$$

$$H_1 = -(i\omega\mu_0)^{-1}(\text{curl curl } F - j_m) + \text{curl } A. \tag{34}$$

From (31)-(34) and (9) one gets

$$E(x) = E_0(x) + \frac{\varepsilon'-\varepsilon_0}{\varepsilon_0}\text{curl curl}\int G(x,y)Edy$$
$$- \frac{\varepsilon'-\varepsilon_0}{\varepsilon_0}\eta E + i\omega(\mu-\mu_0)\text{curl}\int G(x,y)Hdy, \tag{35}$$

$$H = H_0(x) + \frac{\mu-\mu_0}{\mu_0}\text{curl curl}\int_D G(x,y)Hdy$$
$$- \frac{\mu-\mu_0}{\mu_0}H - i\omega(\varepsilon'-\varepsilon_0)\text{curl}\int G(x,y)Edy. \tag{36}$$

The system (35)-(36) can be solved by iterations if

$$\left(\left|\frac{\mu-\mu_0}{\mu_0}\right| + \left|\frac{\varepsilon'-\varepsilon_0}{\varepsilon_0}\right|\right)(1 + k_0^2 a^2) \ll 1. \tag{37}$$

Indeed, under the assumption (37) the norm in $L^2(D)$ of the operator of system (35)-(36) is less than 1. Let us verify this statement. We have

$$\left|\left|\frac{\varepsilon'-\varepsilon_0}{\varepsilon_0}\eta E\right|\right| \leq \left|\frac{\varepsilon'-\varepsilon_0}{\varepsilon_0}\right| ||E||. \tag{38}$$

Here and below $||\cdot||$ denotes the $L^2(D)$ norm and c denotes various constants. Furthermore,

$$||curl \int G(x,y)Edy|| \le c(1 + k_0a)||E||, \tag{39}$$

$$||curl\ curl \int G(x,y)Edy|| \le c(1 + k_0^2a^2)||E||. \tag{40}$$

Inequalities (39) and (4) can be proved as follows. Note that

$$4\pi G(x,y) = \frac{1}{|x-y|} + ik - \frac{k_0^2|x-y|}{2} + O(k_0^3|x-y|^2), \tag{41}$$

if $k_0|x-y| \ll 1$. Hence

$$D^2G = D^2 \frac{1}{|x-y|} + O\left(\frac{k_0^2}{|x-y|}\right). \tag{42}$$

We have

$$\left|\left|\int \frac{Edy}{x-y}\right|\right| \le ||E||\left(\int\int \frac{dxdy}{|x-y|^2}\right)^{\frac{1}{2}} = ||E||O(a^2) \tag{43}$$

$$\left|\left|\int \frac{Edy}{|x-y|}\right|\right|_{W_2^2(D)} \le c||E||. \tag{44}$$

Inequality (44) is known in the theory of elliptic boundary value problems and embedding theorems (see, e.g., [12]). The desired statement follows from the above estimates. One can see that the body can be large ($k_0a \gg 1$) provided that

$$\left|\frac{\varepsilon'-\varepsilon_0}{\varepsilon_0}\right| + \left|\frac{\mu-\mu_0}{\mu_0}\right|$$

is sufficiently small.

Let us set

$$g(n) = \int \exp\{-ik_0(n,y)\}dy, \tag{45}$$

iterate once system (35)-(36) and calculate the scattering amplitude. This yields

$$f(n,k) = -\frac{\varepsilon'-\varepsilon_0}{4\pi\varepsilon_0} k_0^2 g(n)n \times [n,E_0] - g(n) \frac{\omega(\mu-\mu_0)}{\mu_0} n \times H. \tag{46}$$

If D is a ball of radius a, then

$$g(n) = 4\pi a^3 \frac{\sin(k_0a) - k_0a\ \cos\ k_0a}{(k_0a)^3}. \tag{47}$$

If D is a cylinder with radius a and length 2L, then

$$g(n) = 2L \ \frac{\sin(k_0 L \cos \theta)}{k_0 L \cos \theta} \ \frac{J_1(k_0 a \sin \theta)}{k_0 a \sin \theta} \ , \tag{48}$$

where θ is the angle between the axis of the cylinder and the unit vector n, and $J_1(x)$ is the Bessel function.

5. Many-body electromagnetic wave scattering can be developed along the lines of §3.

6. Let us derive the following *formula for the scattering matrix for the electromagnetic wave scattering by a single body under the assumption* (2):

$$S = \frac{k^2 V}{4\pi} \begin{pmatrix} \mu_0 \beta_{11} + \alpha_{22} \cos \theta - \alpha_{32} \sin \theta & \alpha_{21} \cos \theta - \alpha_{31} \sin \theta - \mu_0 \beta_{12} \\ \alpha_{12} - \mu_0 \beta_{21} \cos \theta + \mu_0 \beta_{31} \sin \theta & \alpha_{11} + \mu_0 \beta_{22} \cos \theta - \mu_0 \beta_{32} \sin \theta \end{pmatrix}, \tag{49}$$

where θ is the angle of scattering and β_{ij} and α_{ij} are the polarizability tensors defined in Chapter 5. In Chapter 5 approximate analytical formulas for calculating these tensors are given.

If assumption (1) holds then one can neglect terms involving β_{ij} in (49). Let us prove (49). Let the origin be inside D, the initial field be a plane wave propagating in the positive direction e_3 of the z-axis, n be a unit vector, and θ be the angle between e_3 and n (the angle of scattering). Let E_1, E_2 be the projections of the initial electric field onto the axes OX and OY, and f_1, f_2 be the projections of the scattered electric field onto the axes OX^1, OY^1. The axis OZ^{-1} is assumed to be in the direction of n. The plane (OZ,OY) coincides with the plane (OZ^1,OY^1) and is called the plane of scattering.

The scattering matrix is defined by the formula $f_E = SE$:

$$\begin{pmatrix} f_2 \\ f_1 \end{pmatrix} = \begin{pmatrix} S_2 & S_3 \\ S_4 & S_1 \end{pmatrix} \begin{pmatrix} E_2 \\ E_1 \end{pmatrix} \tag{50}$$

Formula (49) gives this matrix explicitly. All the elements of the S-matrix are calculated by the same method. Let us derive in detail the formula for S_2. Let e_j (e'_j) be the unit vectors of the above coordinate systems. Then $(e'_2,e_1) = 0$, $(e'_2,e_2) = \cos \theta$, $(e'_2,n) = -\sin \theta$, $f_2 = S_2 E_2 + S_3 E_1$. On the other hand,

$$f_2 = (f, e_2') = \frac{k^2}{4\pi\varepsilon_0} ([n,[P,n]],e_2') + \frac{k^2}{4\pi} \left(\frac{\mu_0}{\varepsilon_0}\right)^{\frac{1}{2}} ([M,n],e_2'),$$

where (e,f) $([e,f])$ denotes the scalar (vector) product. We have

$$([n[P,n]],e_2') = (P-n(P,n),e_2') = (P,e_2') = \varepsilon_0 V\alpha_{ij}E_j(e_i,e_2')$$

$$= \varepsilon_0 V\{(\alpha_{21}E_1 + \alpha_{22}E_2)\cos\theta - (\alpha_{31}E_1 + \alpha_{32}E_2)\sin\theta\}, \tag{51}$$

$$([M,n],e_2') = (n,e_2',M) = -(e_1,M) = -\mu_0 V(\beta H, e_1)$$

$$= -\mu_0 V(\beta_{11}H_1 + \beta_{12}H_2) = -\mu_0 V\left(\frac{\varepsilon_0}{\mu_0}\right)^{\frac{1}{2}}(-\beta_{11}E_2 + \beta_{12}E_1), \tag{52}$$

where the formulas

$$H_1 = -\left(\frac{\varepsilon_0}{\mu_0}\right)^{\frac{1}{2}} E_2, \qquad H_2 = \left(\frac{\varepsilon_0}{\mu_0}\right)^{\frac{1}{2}} E_1 \tag{53}$$

were used. From (51) and (52) we find

$$S_2 = \frac{k^2 V}{4\pi}(\alpha_{22}\cos\theta - \alpha_{32}\sin\theta + \mu_0\beta_{11}) \tag{54}$$

as the coefficient of E_2. Formulas for the other elements of the S-matrix can be obtained similarly.

Knowing the S-matrix for a single small body, one can find the refraction index tensor $n_{ij} = \delta_{ij} + 2\pi Nk^{-2}S_{ij}(0)$ of the rarefied medium consisting of many small particles, the coefficient of absorption $x = N\sigma = 4\pi Nk^{-1} \text{Im } S(0)$, the crosssection $\sigma = 2\pi k^{-1} \text{ tr Im } S(0)$ for anisotropic scattering, etc. Here N is the number of the particles per unit volume, tr denotes the trace of a matrix, and Im denotes the imaginary part of a complex number.

§5. Radiation from Small Apertures and the Skin Effect for Thin Wires

1. Let F be an aperture in an infinite conducting plane, α_0 be its coefficient of electrical polarizability, β_{ij}^0, $1 \le i,j \le r$, be its tensor of magnetic polarizability, the x_3-axis be perpendicular to the plane and e_j, $1 \le j \le 3$, be the coordinate unit vectors. We assume that the electric field in the halfspace $x_3 < 0$ is $E_0'e_3$ and in the halfspace x_3 0 the electrostatic potential $\phi \sim (p,x)/(4\pi\varepsilon_0|x|^3)$, $E = -\nabla\phi$. The electric dipole moment P can be calculated from the formula

$$P = \alpha_0\varepsilon_0 E_0'e_3. \tag{1}$$

The magnetic field in the half-space $x_3 < 0$ is $H_0' = H_{01}'e_1 + H_{02}'e_2$ and its asymptotic behavior in the half-space $x_3 > 0$ is given by $\psi \sim (M,x)/(4\pi\mu_0|x|^3)$, where ψ is the magnetostatic potential, M is the magnetic dipole moment, $H = -\nabla\psi$ for $x_3 > 0$, and

$$M_i = \beta_{ij}^0 \mu_0 H'_{0j}. \tag{2}$$

Let $\tilde{\beta}$ and $\tilde{\alpha}_{ij}$ denote the magnetic polarizability coefficient and the electric polarizability tensor of the thin magnetic film and the thin metallic screen with the shape of F. The following theorem is a duality principle in electrostatics.

Theorem 1. *The following formulas hold*

$$\alpha_0 = -\tilde{\beta}/2, \qquad \beta_{ij}^0 = -\tilde{\alpha}_{ij}/2. \tag{3}$$

Remark 1. Formulas for calculating the values of $\tilde{\beta}$ and $\tilde{\alpha}_{ij}$ are given in Chapter 5. If one knows these values, one can find α_0 and β_{ij}^0 from (3) and P and M from (1) and (2). Knowing P and M, one can calculate the radiation from the aperture F from (4.13).

Proof of Theorem 1: Let us formulate two principles:

(A) Let there be an initial electrostatic field $\tilde{E}_0^{(2)} = E_0 e_3$ in the half-space $x_3 < 0$ bounded by the conducting plane $x_3 = 0$. If we cut an aperture F in the plane $x_3 = 0$ then the field $E^{(2)}$ in the half-space $x_3 > 0$ can be calculated from the formula $E^{(2)} = H^{(1)} - H_0^{(1)}$, where $H^{(1)}$ is the magnetic field which is present when a magnetic plate F with $\mu = 0$ is placed in the initial field $H_0^{(1)} = -\frac{1}{2}\tilde{E}_0^{(2)} = -\frac{1}{2}E_0 e_3$.

(B) Let there be a magnetostatic field $H_0^{(2)}$ parallel to the plane $x_3 = 0$ in the half-space $x_3 < 0$ bounded by the plane $x_3 = 0$ with $\mu = 0$. If we cut an aperture F in the plane then the field $H^{(2)}$ in the half-space $x_3 > 0$ can be calculated from the formula $H^{(2)} = -(E^{(1)} - E_0^{(1)})$, where $E^{(1)}$ is the electric field which is present when the metallic plate F is placed in the initial field $E_0^{(1)} = \frac{1}{2}H_0^{(2)}$.

Formula (3) follows immediately from these principles and from the definition of α_0, $\tilde{\beta}$, β_{ij}^0, $\tilde{\alpha}_{ij}$. Both principles can be proved similarly. We give the proof of (A).

Let $S = R^2 \setminus F$. We have $E^{(2)} = -\nabla u$, where

$$u = \begin{cases} \phi, & x_3 > 0, \\ -E_0 x_3 + \phi, & x_3 < 0, \end{cases}$$

$\Delta\phi = 0$ outside S, $\phi|_S = 0$, $\phi(\infty) = 0$, and u, $\partial u/\partial x_3$ are continuous when crossing F, i.e., $(\partial\phi/\partial x_3)_+ = -E_0 + (\partial\phi/\partial x_3)_-$. By symmetry we have $\phi(\hat{x},x_3) = \phi(\hat{x},-x_3)$, $\hat{x} = (x_1,x_2)$. Hence $(\partial\phi/\partial x_3)_- = -(\partial\phi/\partial x_3)_+$, $(\partial\phi/\partial x_3)_+ = -\frac{1}{2}E_0$. Here $(\partial\phi/\partial x_3)$ are the limiting values of $\partial\phi/\partial x_3$ on F for $x_3 \to \pm 0$. So $\Delta\phi = 0$ for $x_3 > 0$, $\phi|_S = 0$, $\phi(\infty) = 0$, $(\partial\phi/\partial x_3)_+ = -\frac{1}{2}E_0$, and $E^{(2)} = -\nabla\phi$ for $x_3 > 0$. The field $H^{(1)} - H_0^{(1)} = -\nabla\psi$ for $x_3 > 0$, where $\Delta\psi = 0$, $\psi(\infty) = 0$, and by symmetry $\psi(\hat{x},-x_3) = -\psi(\hat{x},x_3)$. The magnetostatic potential $v = \frac{1}{2}E_0 x_3 + \psi$ satisfies the condition $(\partial v/\partial N)|_F = 0$, where N is the outward pointing normal to F. Hence $(\partial\psi/\partial x_3)_+ = -\frac{1}{2}E_0$. As ψ is odd in x_3, we conclude that $\psi|_{x_3=0} = 0$, $\psi|_S = 0$. Hence ϕ,ψ are the solutions of the same boundary value problem in the half-space $x_3 > 0$. The solution of this problem is unique. Hence $\phi \equiv \psi$ for $x_3 > 0$. This means that $E^{(2)} = H^{(1)} - H_0^{(1)}$ for $x_3 > 0$. Principle (A) is proved. □

Example. For disk with radius a we have $\tilde{\beta} = -(8/3)a^3$, $\tilde{\alpha} = (16/3)a^3\delta_{ij}$, $1 \leq i,j \leq 2$, $\alpha_0 = (4/3)a^3$, $\beta_{ij}^0 = -(8/3)a^3\delta_{ij}$, $1 \leq i,j \leq 2$, in SI units.

2. In Chapter 5 some two-sided variational estimates of $\tilde{\beta}$ and $\tilde{\alpha}_{ij}$ were given. In the special case in which F is a plane aperture one can give another variational estimate of $\tilde{\beta}$. Actually we will derive the estimate for $\alpha_0 = -\tilde{\beta}/2$.

Let $S = R^2 \setminus F$, be the complement of F in the plane, and let

$$g(s) = \int_F r_{st}^{-1}dt, \quad \alpha = (2\pi)^{-1}\int\int_F r_{st}^{-1} dsdt. \qquad (4)$$

Then the following variational principle holds:

$$\alpha - \alpha_0 = \max \frac{1}{2\pi} \frac{\left(\int_S g(t)u(t)dt\right)^2}{\int_S\int_S \frac{u(s)u(t)}{r_{st}} dsdt}, \qquad (5)$$

where the admissible functions should satisfy the edge condition (1.1.17) and ensure convergence of the integrals in (5). Principle (5) allows one to obtain some upper bounds for α_0.

Let us derive (5). Let $E_0' = E_0 e_3$ be the electric field in the half space $x_3 < 0$ and the aperture F is cut in the conducting plane $x_3 = 0$. Then the potential ϕ in the half space $x_3 > 0$ can be written as

$$\phi(x) = 2\int_F \phi(t)\ \frac{\partial G_0}{\partial x_3}\ dt, \qquad x_3 > 0, \tag{6}$$

where

$$G_0(x,y) = (4\pi r_{xy})^{-1}, \tag{7}$$

and

$$\phi(x) = -2\int_F \phi(t)\ \frac{\partial G}{\partial x_3}\ dt - E_0 x_3, \qquad x_3 < 0. \tag{8}$$

The potential $\phi(x)$ and its derivatives are continuous when crossing the aperture F and

$$\phi\big|_S = 0. \tag{9}$$

When $|x| \to \infty$, $x_3 > 0$ one has

$$\phi(x) \sim \frac{2\varepsilon_0 \int_F \phi(t) dt\ x_3}{4\pi\varepsilon_0 |x|^3} = \frac{(P,x)}{4\pi\varepsilon_0 |x|^3}, \tag{10}$$

where

$$P = \alpha_0 \varepsilon_0 E_0 e_3, \tag{11}$$

and

$$\alpha_0 = \frac{2}{E_0} \int_F \phi(t) dt. \tag{12}$$

Let σ denote the charge density on S.

$$\sigma = -\varepsilon_0 \frac{\partial\phi}{\partial x_3}\bigg|_{x_3=+0}. \tag{13}$$

Green's formula implies

$$\phi(x) = \int_{x_3=0}\left(\phi(t)\ \frac{\partial G_0(x,t)}{\partial x_3} - G_0(x,t)\ \frac{\partial\phi}{\partial x_3}\right) dt. \tag{14}$$

From (9), (13), and (14) it follows that

$$\phi(x) = \int_F \frac{\partial G_0}{\partial x_3}\ dt + \frac{1}{\varepsilon_0} \int_S G_0(x,t)\sigma dt - \int_F G_0(x,t)\ \frac{\partial\phi}{\partial x_3}\ dt. \tag{15}$$

Let us show that

$$\frac{\partial\phi}{\partial x_3}\bigg|_F = -\frac{E_0}{2}. \tag{16}$$

This follows from (6), (8), and the condition

$$\left(\frac{\partial \phi}{\partial x_3}\right)\Big|_{x_3=+0} = \left(\frac{\partial \phi}{\partial x_3}\right)\Big|_{x_3=-0} \quad \text{on} \quad F. \tag{17}$$

Let us take $x \in S$ in (15) and take into account (16). This yields

$$\int_S \frac{\sigma(t)dt}{r_{st}} = -\frac{\varepsilon_0 E_0}{2} g(s), \quad s \in S, \tag{18}$$

where $g(s)$ is defined in (4). Let $x \to s \in F$, $x_3 > 0$ in (15). This yields

$$\phi(s) = \frac{\phi(s)}{2} + \frac{1}{\varepsilon_0} \int_S G_0(s,t)\sigma dt + \frac{E_0}{8\pi} g(s), \tag{19}$$

which is equivalent to the equation

$$\phi(s) = \frac{1}{2\pi\varepsilon_0} \int_S \frac{\sigma(t)dt}{r_{st}} + \frac{E_0}{4\pi} g(s). \tag{20}$$

From (20) and (12) it follows that

$$\alpha_0 = \frac{1}{\pi\varepsilon_0 E_0} \int_S \sigma(t)g(t)dt + \alpha, \tag{21}$$

where α is defined in (4). This can be written as

$$\alpha - \alpha_0 = -\frac{1}{\pi\varepsilon_0 E_0} \int_S \sigma(t)g(t)dt. \tag{22}$$

From (18), (22), and Theorem 3.2.1 formula (5) follows. In the derivation of (5) we used some ideas from [5].

3. Consider the skin effect in thin wires. Let the axis of the wire be directed along the x_3-axis Γ be the boundary of the cross section of the wire, a be the diameter of Γ, $ka \ll 1$. One can consider also wires the axes of which are curves with radius of curvature $R \gg a$. We assume that $\delta \ll a$ where δ is the skin depth defined in §4. Let L be the length of Γ, J be the total current in the wire, and

$$Aj = \frac{1}{\pi} \int_\Gamma \frac{\partial}{\partial N_s} \ln \frac{1}{r_{st}} j(t)dt, \tag{23}$$

where N_s is the unit outward pointing normal to Γ at the point s.

Proposition 1. *Under the above assumptions the current distribution on Γ can be found by the iterative process*

$$j_{n+1} = -Aj_n, \quad j_0 = \frac{J}{L}, \quad j = \lim_{n \to \infty} j_n(t). \tag{24}$$

Proof: It is sufficient to note that under the above assumptions the problem about the current distribution on Γ can be formulated as follows. Let $v(x_1,x_2)e_3$ be the vector potential of the static magnetic field corresponding to the current J. Then

$$\frac{\partial^2 v}{\partial x_1^2} + \frac{\partial^2 v}{\partial x_2^2} = 0 \quad \text{in } \Omega, \qquad v\big|_\Gamma = \text{const}, \tag{25}$$

$$v \sim \frac{\mu_0 J}{2\pi} \ln \frac{1}{r} \quad \text{as} \quad r = (x_1^2 + x_2^2)^{\frac{1}{2}} \to \infty, \tag{26}$$

$$-\frac{1}{\mu_0} \frac{\partial v}{\partial N}\bigg|_\Gamma = j(t), \qquad J = \int_\Gamma j(t)\,dt. \tag{27}$$

If we look for the solution of the problem (25)-(27) of the form

$$v(x) = \frac{\mu_0}{2\pi} \int_\Gamma \ln \frac{1}{r_{xt}} j(t)\,dt, \tag{28}$$

then from (27) it follows that

$$j = -\frac{Aj - j}{2}$$

or

$$j = -Aj. \tag{29}$$

Proposition 1 follows now from Theorem 6.1.2. □

§6. The Inverse Problem of Radiation Theory

1. Suppose that we are interested in measuring the electromagnetic field in the aperture of the mirror antenna. A possible method for making such measurements is as follows. Let us assume that the wavelength range is $\lambda \sim 3$ cm and let us place at some point x_0 in the aperture of the antenna a small probe of dimension a, $ka \ll 1$, $k = 2\pi\lambda^{-1}$. Let E_0, H_0 denote the electromagnetic field at the point x_0 and E, H denote the field scattered by the probe in the far-field zone. Note that for a small probe the far-field zone which is defined by the known condition $ka^2 r^{-1} \ll 1$ is in fact close to the probe. For example, if $\lambda = 3$ cm, $a = 0.3$ cm then $ka^2 = 0.19$ cm. Therefore if $r = 2$ cm then $ka^2 r^{-1} \approx 0.1 \ll 1$. Let us assume for simplicity that the probe material is such that the magnetic dipole radiation from the probe is negligible. In this case the electric field scattered by the probe in the direction n can be calculated from the formula (4.13) as

$$E = \frac{k^2}{4\pi\varepsilon_0} [n,[P,n]], \tag{1}$$

where

$$P_i = \alpha_{ij}(\gamma)\varepsilon_0 VE_{0j}, \quad \gamma = \frac{\varepsilon'-\varepsilon_0}{\varepsilon'+\varepsilon_0}. \tag{2}$$

Here V is the volume of the probe, ε_0 is its dielectric constant, $\alpha_{ij}(\gamma)$ is its electric polarizability tensor, k is the wave number of the field in the aperture, E_0 is the electric field at the point x_0 where the probe was placed, and n is the unit vector. Let n_1 and n_2 be two noncollinear unit vectors, and E_j, $j = 1,2$, be the scattered fields corresponding to n_j. We will solve the following

Problem. *Find* E_0, H_0 *from the measured* E_j, $j = 1,2$. We assume that the tensor $\alpha_{ij}(\gamma)$ is known. In Chapter 5 some explicit analytical approximate formulas for $\alpha_{ij}(\gamma)$ are given. From (1) it follows that

$$E_j = b\{P-n_j(P,n_j)\}, \quad b = \frac{k^2}{4\pi\varepsilon_0}, \quad j = 1,2. \tag{3}$$

Therefore

$$bP = E_1 + bn_1(P,n_1) = E_2 + bn_2(P,n_2). \tag{4}$$

Let us choose for simplicity n_1 perpendicular to n_2. Then it follows from (4) that

$$b(P,n_2) = (E_1,n_2), \tag{5}$$

$$b(P,n_1) = (E_2,n_1). \tag{6}$$

Therefore

$$P = b^{-1}E_1 + b^{-1}n_1(E_2,n_1) = b^{-1}E_2 + b^{-1}n_2(E_1,n_2). \tag{7}$$

Thus one can find vector P from the knowledge of E_1 and E_2. If P is known then E_0 can be found from the linear system

$$\alpha_{ij}(\gamma)\varepsilon_0 VE_{0j} = P_i, \quad 1 \le i \le 3. \tag{8}$$

The matrix of this system is positive definite because the tensor α_{ij} has this property (see Chapter 5). (This follows also from the fact that $\frac{1}{2}\alpha_{ij}\varepsilon_0 VE_{0j}E_{0i}$ is the energy of the dipole P in the field E_0.) Therefore the system (8) can be uniquely solved for E_{0j}, $1 \le j \le 3$. We proved that the above Problem has a unique solution and gave a simple algorithm for the solution of this problem. The key point in the above argument is the fact that the matrix $\alpha_{ij}(\gamma)$ is known explicitly (from Chapter 5).

2. In applications the problem of finding the distribution of particles according to their sizes is often of interest. Suppose that there is
a medium consisting of many particles and the condition (4.3) is satisfied.
We assume that the medium is rarefied, i.e., $d \gg a$, where a is the
characteristic dimension of the particles. Let us assume for simplicity
that the particles are spherical. Then the scattering amplitude for a
single particle can be calculated from formulas (4.46) and (4.47). The
scattering amplitude is the function $f(n,k,r)$ of the radius r of the
particle. Suppose that $\phi(r)$ is the density of the distribution of the
particles according to their sizes, so that $\phi(r)dr$ is the number of the
particles per unit volume with the radius in the interval $(r,r+dr)$.
Then the total scattered field in the direction n can be calculated
from the formula

$$F(n,k) = \int_0^\infty \phi(r)f(n,k,r)dr. \tag{9}$$

Let us assume that we can measure $F(n,k)$ for a fixed k and all direc-
tions n. Then (9) can be considered as an integral equation of the
first kind for an unknown function $\phi(r)$.

3. Suppose that we can measure the electric field scattered by a
small particle ($ka \ll 1$) of an unknown shape. The initial field we denote by E_{0j}, the scattered field by f_j. Let us assume that the magnetic dipole radiation is negligible. The problem is to find the shape
of the small particle.

First let us note that every small particle scatters electromagnetic
wave like some ellipsoid. Indeed, the main term in the scattered field
is the dipole scattering. We have seen in Subsection 1 that the knowledge
of the scattered field allows one to find the dipole moment P and that
equation (2) holds. This equation allows one to find the $\alpha_{ij}(\gamma)$ corresponding to the particle. This tensor is determined if one knows its
diagonal form. Let $\alpha_1, \alpha_2, \alpha_3$ be the eigenvalues of the tensor $\alpha_{ij}(\gamma)$.
Then an ellipsoid with the semiaxes proportional to α_j scatters as the
above body. Therefore one can identify the shape of the small scatterer
by giving the three numbers $(\alpha_1,\alpha_2,\alpha_3)$. These numbers are the eigenvalues of the tensor $\alpha_{ij}(\gamma)$ which can be calculated from the known initial field E_{0j} and the measured scattered field f_i. For example, one
can take $E_{0j} = \delta_{ij}$. Then $P_i = \alpha_{ij}(\gamma)V\varepsilon_0$. We assume that the particle
is homogeneous and its dielectric constant ε is known, so that γ in
(2) is known. For an ellipoid the polarizability tensor in the diagonal

form is $\alpha_{ij} = \alpha_j \delta_{ij}$, where $\alpha_j = (\varepsilon' - \varepsilon_0)(\varepsilon_0 + (\varepsilon' - \varepsilon_0)n^{(j)})^{-1}$, where ε' is the dielectric constant of the ellipsoid and $n^{(j)}$ are the depolarization coefficients. These coefficients are calculated explicitly with the help of the elliptic integrals and they are tabulated in [17].

Problems

1. Write a computer program for calculating the capacitance and the
   polarizability tensors for a body of arbitrary shape. According to
   formulas (3.1.12) and (5.1.9)-(5.1.13), the program will calculate
   multiple integrals over the surface of the body. The integrands are
   functions with weak singularities, e.g.,

   $$\int_\Gamma \int_\Gamma r_{st}^{-1} ds dt, \quad \int_\Gamma \int_\Gamma r_{st}^{-1} N_i(s) N_j(t) ds dt,$$

   $$\int_\Gamma \int_\Gamma r_{st}^{-1} \left(\int_\Gamma \frac{\partial r_{tt_1}^{-1}}{\partial N_t} dt_1 \right) ds dt.$$

   Finding good algorithms for calculating multiple integrals of functions with moving weak singularities is a problem of considerable
   general interest.

2. Carry out a numerical study of the dependence of the scattering amplitude on: (1) the shape of the body, (2) on the boundary conditions, (3) on the coating of the body (e.g., a flaky-homogeneous body
   with two layers of which the exterior layer is thin).

3. Carry out a numerical study of the many-body problem using formulas
   (7.3.11), (7.3.14), (7.3.23), (7.3.24), (7.3.27), (7.3.31), and
   (7.3.32).

4. Study the inverse problem of finding the properties of the medium
   consisting of many small particles from the scattering data.

5. Develop a theory of elastic wave scattering by small bodies of arbitrary shape similar to the theory of acoustic and electromagnetic
   wave scattering given in Chapter 7.

Bibliographical Notes

The basic results mentioned in the introduction are presented in the
books [2], [3], [10], [17], [18], [21], [22], [24], [32]. Variational
principles and two-sided estimates of various functionals of static fields
are given in [27], [25]. Low-frequency scattering, first studied by
Rayleigh (1871), is studied in [29], [31], [17], [9], [13], [14], [11],
[8]. Scattering from small holes is studied in [1], [19], [5]. Potential theory for domains with smooth boundaries is given in [7]. The theory
for domains with nonsmooth boundaries is presented in [3], [15], and in
the paper by E. B. Fabes, M. Jodeit, Jr., and N. Riviere, Potential techniques for boundary value problem on C^1 domains, *Acta Math.*, *141*, (1978),
165-186. Reference material and an extensive bibliography on electrical
capacitance can be found in [10]. There is an extensive literature on
scattering by a system of many bodies and wave propagation in random media,
topics outside of the scope of this book. Among many contributors to this
field are L. Foldy, The multiple scattering of waves, *Phys. Rev.*, *67*,
(1945), 107-119; M. Lax, Multiple scattering of waves, *Rev. Mod. Phys.*,
23, (1951), 287-310; The effective field in sense systems, *Phys. Rev.*, *88*,
(1952), 621-629; J. Keller, Wave propagation in random media, *Proc. Symp.
Appl. Math.*, *13*, 227-246, AMS, Providence, R. I., 1962; Stochastic equations and Wave Propagation in Random Media, ibid., *16*, (1964), 145-170;
V. Twersky, Propagation in pair-correlated distributions of small-spaced
lossy scatterers, *J. Opt. Soc. Amer.*, *69*, (1979), 1567-1572; Coherent
scalar field in pair-correlated random distribution of aligned scatterers,
J. Math. Phys., *18*, (1977), 2468-2486; P. Waterman, J. Korringa, S. Ström,
and V. Varadan in *Acoustic, Electromagnetic and Elastic Wave Scattering-
Focus on the T-matrix Approach*, Pergamon Press, N.Y., 1980; Yu. Kravtsov,

S. Rytov, and V. Tatarskij, Statistical problems in diffraction theory,
Soviet Phys. Uspekhi, 18, (1975), 118-130; Yu. Barabanenkov, Yu. Kravtsov,
S. Rytov, and V. Tatarskij, State of the theory of wave propagation in
randomly nonhomogeneous medium, ibid., *13,* (1971), 551; V. Finkelberg,
Wave propagation in random media, *JETP, 53,* (1967), 401-415; V. Tatarskij,
Wave propagation in a turbulent atmosphere, Nauka, M., 1967. V. Marchenko,
E. Hruslov, Boundary value problems in domains with granular boundary,
Naukova, Kiev, 1974. This list is incomplete.

Integral equations of the first kind were used in electrostatics
[34]. In [4] an integral equation of the second kind was derived for
screens (non-closed surfaces). A numerical approach to problem (6.3.61)
differing from the one given in §6.3 was given in [36]. A recent paper
by P. Colli Franzone and E. Magenes, On the inverse potential problem of
electrocardiology, *Calcolo, 4,* (1980), 459-538 discusses the possibility
of calculating the cardiac electric potential of a human body from the
surface measure potential. A computer program for calculating the elements of the polarizability tensor of rotationally symmetric metallic
bodies was given in T. B. A. Senior and D. J. Ahlgren, Rayleigh scattering, *IEEE Trans., AP-21,* (1973), 134.

The main results presented in this book were obtained by the author
[28]. These results include: (1) approximate analytical formulas for
polarizability tensors and capacitances, (2) two-sided estimates of the
polarizability tensors, (3) approximate analytical formulas for the scattering amplitude and scattering matrix in the problem of wave scattering
by a small body of an arbitrary shape and by a system of such bodies,
(4) investigation of the influence of the boundary conditions on the scattering amplitude.

In [28u, p. 144] the factor 2 in the right-hand side of formula (6.20)
was omitted by an oversight. Therefore this factor is not present in some
other formulas on p. 144-146. In §7.3 of this book the corresponding formulas are given with corrections.

Bibliography

[1]   Bethe, H. Theory of diffraction by small holes, *Phys. Rev., 66,*
      (1944), 163-182.

[2]   Buhgolz, G. *Calculating of Electric and Magnetic Fields,* IL,
      Moscow, 1961.

[3]   Burago, Yu. and Maz'ya, V. *Potential Theory and Function Theory
      for Irregular Regions,* Consult. Bureau, N.Y., 1969.

[4]   Feld, Ya. and Suharevskij, I. Reduction of diffraction problems
      on non-closed surfaces to integral equations of the second kind,
      Radio Engrg. Electron. Phys., 11, (1966), 1017-1024.

[5]   Fihmanas, R. and Fridberg, P. Two-sided estimates for the coeff-
      icients of polarizability in diffraction from small holes, *Soviet
      Phys. Dokl., 189,* (1969), 969-972.

[6]   Greenberg, G. A. *Selected Topics in the Mathematical Theory of
      Electrical and Magnetical Phenomena,* Ac. Sci. USSR, Leningrad,
      1948 (Russian).

[7]   Günter, N. *Potential Theory and Its Applications to Basic Problems
      of Mathematical Physics,* Ungar, N.Y., 1967.

[8]   Hönl, G., Maue, A. and Westpfahl, K. *Theorie der Beugung,* Springer-
      Verlag, Berlin, 1961.

[9]   Hulst, Van de. *Light Scattering by Small Particles,* Mir, Moscow,
      1961 (Russian).

[10] Jossel, Ju., Kochanov, E. and Strunskij, M. *Calculation of Elec-
      trical Capacity,* Energija, Leningrad, 1969 (Russian).

[11] Jones, D. S. Low frequency electromagnetic radiation, *J. Inst.
      Math. Appl., 23,* (1979), 421-427.

[12] Kantorovich, L. and Akilov, G. *Functional Analysis,* Macmillan,
      N.Y., 1964.

[13] Keller, J., et al. Dipole moments in Rayleigh scattering, *J. Inst.
      Appl. Math., 9,* (1972), 14-22.

[14] Kleinman, R. Low frequency electromagnetic scattering, in *Electro-
      magnetic Scattering,* ed. P. Uslenghi, Acad. Press, N.Y., 1978.

[15]   Kral, I. *Integral Operators in Potential Theory,* Springer-Verlag, N.Y., 1980.

[16]   Krasnoselskij, M., Vainikko, G., Zabreiko, P., Rutickij, Ja., and Stecenko, V. *Approximate Solution of Nonlinear Equations,* Wolters-Noordhoff, Groningen, 1972 (1969), MR 27#4271.

[17]   Landau, L. and Lifschitz, E. *Electrodynamics of Continuous Media,* Pergamon Press, N.Y., 1960.

[18]   Lebedev, N. et al. *Collection of Problems in Mathematical Physics,* GITTL, Moscow, 1955 (Russian).

[19]   Levine, H. and Schwinger, J. On the theory of electromagnetic wave diffraction by an aperture in an infinite plane conducting screen, *Comm. Pure Appl. Math., 3,* (1950), 355.

[20]   Mihlin, S. *Variational Methods in Mathematical Physics,* Macmillan, N.Y., 1964, MR 30#2712.

[21]   Miroljubov, N. Methods of calculating of electrostatic fields, High School, Moscow, 1963 (Russian).

[22]   Morse, P. and Feshbach, M. *Methods of Theoretical Physics,* vols. 1 and 2, McGraw-Hill, N.Y., 1953.

[23]   Mushelishvili, N. *Singular Integral Equations,* Noordhoff Int., Leyden, 1972.

[24]   Newton, R. *Scattering of Waves and Particles,* McGraw-Hill, N.Y., 1966.

[25]   Noble, B. *Wiener-Hopf Method for Solution of Partial Differential Equations,* Pergamon Press, N.Y., 1958.

[26]   Payne, L. Isoperimetric inequalities and their application, *SIAM Rev., 9,* (1967), 453-488.

[27]   Polya, G. and Szego, G. *Isoperimetric Inequalities in Mathematical Physics,* Princeton Univ. Press, Princeton, 1951.

[28]   Ramm, A. G.

   a.   Iterative solution of the integral equation in potential theory, *Doklady, 186,* (1969), 62-65, MR 41#9462.

   b.   Approximate formulas for tensor polarizability and capacitance of bodies of arbitrary shape and its applications, *Doklady, 195,* (1970), 1303-1306, MR 55#1947; English translation, *15,* (1971), 1108-1111. (Doklady = Sov. Phys. Doklady).

   c.   Calculation of the initial field from scattering amplitude, *Radio Engrg. Electron. Phys., 16,* (1971), 554-556.

   d.   Approximate formulas for polarizability tensor and capacitance for bodies of an arbitrary shape, *Radiofisika, 14,* (1971), 613-620, MR 47#1386.

   e.   Iterative methods for solving some heat transfer problems, *Eng.-Phys. Jour., 20,* (1971), 936-937.

   f.   Electromagnetic wave scattering by small bodies of an arbitrary shape, *Proc. of 5th All-Union Sympos. on Wave Diffraction,* Trudy Mat. Inst. Steklov, Leningrad, 1971, 176-186.

   g.   Calculation of the magnetization of thin films, *Microelectronics, 6,* (1971), 65-68 (with N. Frolov).

h. Calculation of the scattering amplitude of electromagnetic waves by small bodies of an arbitrary shape II, *Radiofisika, 14,* (1971), 1458-1460.

i. Electromagnetic wave scattering by small bodies of an arbitrary shape and relative topics, *Proc. Intern. Sympos. URSI,* Moscow, 1971, 536-540.

j. Calculation of the capacitance of a parallelepiped, *Electricity, 5,* (1972), 90-91 (with Golubkova, Usoskin).

k. On the skin-effect theory, *J. Techn. Phys., 42,* (1972), 1316-1317.

l. Calculation of the capacitance of a conductor placed in aniso-tropic inhomogeneous dielectric, *Radiofisika, 15,* (1972), 1268-1270, MR 47#2284.

m. A remark on integral equations theory, *Differential Equations, 8,* (1972), 1517-1520, MR 47#2284; English translation, 1177-1180.

n. Iterative process to solve the third boundary value problem, *Ibid., 9,* (1973), 2075-2079, MR 48#6861.

o. Light scattering matrix for small particles of an arbitrary shape, *Opt. and Spectroscopy, 37,* (1974), 125-129.

p. Scalar scattering by the set of small bodies of arbitrary shape, *Radiofisika, 17,* (1974), 1062-1068.

q. New methods of calculating the static and quasistatic electro-magnetic waves, *Proc. 5th Intern. Sympos. "Radioelectronics-74",* Sofia, *3,* (1974), 1-8 (report 12).

r. Estimates of some functionals in quasistatic electrodynamics, *Ukrain. Phys. Journ., 5,* (1975), 534-543, MR 56#14165.

s. Boundary value problem with discontinuous boundary condition, *Differential Equations, 13,* (1976), 931-933, MR 54#10830.

t. Wave scattering by small particles, *Optics and Spectrocopy, 43,* (1977), 523-532.

u. *Theory and Applications of Some New Classes of Integral Equations,* Springer-Verlag, N.Y., 1980.

[29] Rayleigh, J. *Scientific Papers,* Cambridge, 1922 (in particular papers from *Phil. Mag.,* vols. 35, 41, 44).

[30] Resvyh, K. *Calculating the Electrostatic Fields,* Energy, Moscow, 1967 (Russian).

[31] Stevenson, A. Solution of electromagnetic scattering problems as power series in the ratio (dimension of scatterer/wavelength), *J. Appl. Phys., 24,* (1953), 1134-1142.

[32] Smythe, W. *Static and Dynamic Electricity,* McGraw-Hill, N.Y., 1939.

[33] Tosoni, O. *Calculation of Electromagnetic Fields on Computers,* Technika, Kiev, 1967 (Russian).

[34] Tsyrlin, L. On a method of solving of integral equations of the first kind in potential theory problems, *J. Vyčisl. Math. and Math. Phys., 9,* (1969), 235-238.

[35]   Wainstein, L. Static problems for circular hollow cylinder of
        finite length, *J. Tech. Phys.*, *32*, (1962), 1165-1173; *37*, (1967),
        1181-1188.

[36]   Wendland, W. et al. On the integral equation method for the plane
        mixed boundary value problem of the Laplacian, *Math. Meth. in the
        Appl. Sci.*, *1*, (1979), 265-321.

[37]   Zabreiko, P., Koshelev, A., Krasnoselskij, M., Mihlin, S.,
        Rakovščik, L. and Stecenko, V. *Integral Equations,* Noordhoff Int.,
        Leiden, 1975.

[38]   Krylov, V. and Shulgina, L. *Reference Book in Numerical Integra-
        tion,* Nauka, Moscow, 1968 (Russian).

List of Symbols

$A \times B = [A,b]$ vector product

$A \cdot B = (A,b)$ scalar product

Λ - Laplacian

∇ - gradient, $\hat{\nabla}$ - surface gradient

tr - trace

$\alpha_{ij}(\gamma)$ - polarizability tensor   §5.1

β_{ij} - magnetic polarizability tensor   §5.1

$\tilde{\alpha}_{ij}$, $\tilde{\beta}_{ij}$ - polarizability tensors for screens and films   §5.4

P_i - electric dipole moment   §5.1

M_i - magnetic dipole moment   §5.4

V - volume of the domain D

S - area of the surface

$r_{st} = |s-t|$

$\psi(t,x) = \dfrac{\partial}{\partial N_t} \dfrac{1}{r_{st}}$

$\dfrac{\partial}{\partial N_{i \atop e}}$ - the limit value of the normal derivative from the interior (exterior)

Γ - closed surface

F - screen (unclosed surface)

L - the edge of F

$D_e = \Omega$ - exterior domain

f_E - scattering amplitude

C_{ij} - electrical inductance coefficients §3.5

$C_{ij}^{(-1)}$ - potential coefficients §3.5

Y_{mn} - the spherical harmonics

ε,μ - dielectric and magnetic constants of the scatterer

$\varepsilon_e = \varepsilon_0$, ε_0 - dielectric and magnetic constants of the medium

$\gamma = (\varepsilon-\varepsilon_0)(\varepsilon+\varepsilon_0)^{-1}$

$G^{\perp}$ - orthogonal complement to the subspace G §7.1

$R(A)$ - range of the linear operator A §7.1

$N(A) = \{f: Af = 0\}$ - the null space of the operator A

st - stationary value

$H_q = W_2^q$ - the Sobolev space §6.3

$\in$ - member of

For some symbols we do not give the page numbers because these symbols are standard.

Theory and Applications of Some New Classes of Integral Equations

by Alexander G. Ramm

Focusing on new results, methods, and applications, **Theory and Applications of Some New Classes of Integral Equations** examines various classes of integral and operator equations important in signal estimation, and filtering, network theory, and wave scattering theory. Topics covered include: solving the multidimensional integral equation in random fields filtering theory, analytical approximation formulas for calculating the scattering matrix for small bodies of arbitrary shape, and calculating the stationary regimes in passive nonlinear networks. Theories and formulas for calculations are given.